Soil Erosion and Reservoir
Sedimentation in Lesotho

Abstract

Chakela, Q. K., 1981: Soil Erosion and Reservoir Sedimentation in Lesotho. *Department of Physical Geography, Uppsala University.* UNGI Rapport No. 54 and Scandinavian Institute of African Studies, Uppsala 1981. 150 pp. ISBN 91-7106-186-X.

The purpose of this study was to document the types, rates and extent of soil erosion and sedimentation within the Roma Valley/Maliele and Khomo-khoana catchments in Lesotho. The drainage areas studied range from 0.2 to 57 km². All are located within lowlands and foothills regions.

The methods used are: (1) reservoir sedimentation surveys, (2) catchment erosion surveys, and (3) measurement of water- and suspended-sediment discharge.

The rates of reservoir sedimentation vary from 0 to 25 cm m^{-2} y^{-1}. These rates correspond to sediment yields of 0—1800 t km^{-2} y^{-1}. The suspended sediment loads range from 270—1400 t km^{-2} y^{-1}. The present rates of gully growth (headward extension) vary from a few decimetres to about a metre per year for the majority of the presently active gullies. Maximum rates of up to 10 m per year were observed in some gullies within the Roma Valley catchment.

The erosion processes vary from landform to landform. On uncultivated mountain slopes, these processes include surface wash (sheet erosion), rainsplash and regolith stripping. Surface wash, rill formation, and wind erosion are active on the cultivated land on the mountain slopes, on the undulating and rolling dissected plains in the foothills and in the lowlands. Mass movements, gully erosion, rill formation, surface wash and rainsplash occur on escarpments and on the overgrazed scree slopes. Gully-, pipe- and channel-erosion predominate along the major streams and rivers, and on the valley-side slopes nearest the main streams.

Besides reservoir sedimentation, deposition occurs also in three other zones: at the foot of the scree slopes where infertile sediments often flood the bottom lands; in the gently-sloping sections of the main stream channels and major gullies; and where gully-side slumps have fallen into the gullies or streams.

Q. K., Chakela, Department of Physical Geography, Uppsala University, P. O. Box 554, S—751 22 Uppsala, Sweden.

Soil Erosion and Reservoir Sedimentation in Lesotho

Qalabane K. Chakela

Scandinavian Institute of African Studies, Uppsala 1981
UNGI Rapport Nr 54
Department of Physical Geography, University of Uppsala

This study also appears as *UNGI Rapport Nr 54* from the Department of Physical Geography, Uppsala University, P.O. Box 554, S-751 22 Uppsala, Sweden.

© Qalabane K. Chakela 1981
ISBN 91-7106-186-X
Printed by Bohusläningens AB, Uddevalla 1981

Preface

The studies presented here were carried out at the Department of Physical Geography, University of Uppsala from 1973 to 1980, under the leadership of Professor Åke Sundborg. My main advisors have been Professor Anders Rapp (now at the University of Lund) who suggested the studies and participated in the selection of the study areas in Lesotho, Drs. Valter Axelsson and Lennart Strömquist.

The aims of these studies were to document the types, rates and extent of different erosion and sedimentation processes active within some selected catchment areas in Lesotho. The information obtained through these studies is hoped to improve the understanding of soil erosion problems facing Lesotho and to supply some basic data to enable land use and land management planning.

The field studies in Lesotho were carried out between 1973 and 1977, and covered the rainy seasons: 1973/74 (3 months), 1974/75 (5 months), 1975/76 (3 months), and 1976/77 (6 months).

During the field work I received assistance from several institutions and persons in Lesotho. The Department of Geography, the National University of Lesotho (N.U.L.), supplied working space and the head of the Department, Professor Gerard Schmitz, acted as my local supervisor and took part in the reconnaissance surveys for the selection of the study areas. The Department of Chemistry, N.U.L., helped me with the analysis of some of the water samples for sediment concentration. The Department of Works and Maintainance, N.U.L., helped me in solving the problems of transport and gave permission to use the University water

intake reservoir for studies. The Bursar's Office, N.U.L., helped me with the arrangements for the payment of my field assistants. The Department of Hydrological and Meteorological Services, Ministry of Water, Energy and Mining, lent me a boat and some hydrometeorological instruments. The Director of the Roma Valley Agricultural project granted me the permission to use the project reservoirs for study. The Chief Surveyor, Department of Lands, Surveys and Physical Planning, Maseru, granted the permission to purchase and publish the aerial photographs over the studied areas. I am deeply grateful for the assistance offered by the abovementioned institutions and persons, and to all those poeple in Lesotho who helped me in one way or another during the course of the field work.

My deepest gratitude goes to all the people at the Department of Physical Geography, University of Uppsala for their support and various contributions during the course of these studies. Maps and diagrams were drawn by Kjerstin Anderson, the photographic reproductions were done by Assar Lindberg. Laila Bagge analized the sediment samples. Karin Fjällström and Kirsti Mesiniemi typed the manuscript. Nigel Rollison and Garnet Williams kindly helped me with the linguistic revision. I acknowledge the help offered by all these people.

Financial support for these studies has been provided by the following funds and institutions: the Faculty of Mathematics and Natural Sciences of the University of Uppsala, the Scandinavian Institute of African Studies, the Secretariat for International Ecology (SIES), the Swedish

Society for Anthropology and Geography (André and Vega Funds), and the Swedish Agency for Research Cooperation with Developing Countries (SAREC).

A final word of gratitude goes to my family: Themba, Lebohang and Birgitta who supported me in so many ways, and 'Mamapele who always managed to solve my accommodation problems in Lesotho during the field periods.

Uppsala February 1981

Qalabane K. Chakela

Contents

1. Introduction

1.1 Aims and presentation of the key problems

Soil erosion and soil conservation have long been major issues in a number of semi-arid areas in Africa, including Lesotho. The main arguments in debate on these problems in Lesotho have been based on only superficial specific information on types of processes involved, their relative importance and their speeds. The present report presents results obtained from a study on the rates of erosion and reservoir sedimentation within some selected catchments in Lesotho. The studies were initiated in October 1973, after a field reconnaissance tour and discussions with Lesotho authorities engaged in natural resources of Lesotho by Prof. A. Rapp, then at University of Uppsala, Prof. G. Schmitz, National University of Lesotho, and the author.

The main purpose of the studies was to collect and evaluate data and observations in order to document types, extent and rates of present day erosion and sedimentation processes in three catchment areas, through detailed studies of a number of small (0.5—30 km²) selected catchment areas.

The problems approached in this study are the following:

1. Trap efficiency of the reservoirs, and rates of sedimentation in reservoirs, expressed in centimeters per year.

2. Types of reservoir sediments and the possible continued use of sediment-filled reservoirs.

3. Importance of surface runoff as compared to subsurface flow.

4. Importance of sheet erosion as compared to gully erosion.

5. Importance of subsurface piping as compared to surface flow in initiation of gullies.

6. Influence of vegetation and soil cover, land use, and slope on the rates and types of erosion and sedimentation processes.

7. Sediment and water discharge from the catchment areas.

The problem of soil erosion in Lesotho manifests itself in three main ways:

1. Loss of arable land through formation of rills and gullies, removal of topsoil by rill and surface wash leading to rapid filling of reservoirs and covering arable land with low fertility sediments.

2. Loss of grazing lands through overgrazing and formation of gullies.

3. Loss of water by rapid runoff after heavy rainstorms where vegetation has been depleted, and excessive drainage by gullies.

The results of these studies are hoped to form and supply improved understanding of the problems and therefore enable a better and more ecologically adapted land use which may lead to conservation and improvement of soil and water resources of Lesotho. Water and soil resources are the largest known assets of the country.

1.2 Location of study areas

The study areas are located within western lowlands and foothills regions of Lesotho and have altitudes ranging from just above 1 500 m to 2 500 m above m.s.l.

The study areas were selected on the basis of availability of background ma-

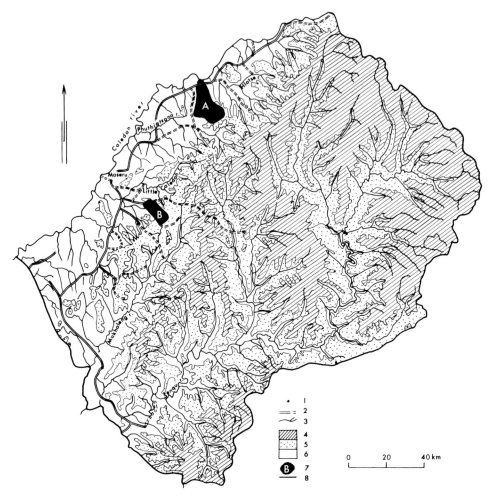

Fig. 1.1. Map of Lesotho, showing relief, hydrography, and the location of investigated catchments. **1**. District headquarters, **2**. Roads, **3**. Hydrography, **4**. Land surface above 8,000 feet, **5**. Land surface between 6,000 and 8,000 feet, **6**. Land surface below 6,000 feet, **7**. Location of the two investigated catchments: A. Khomo-khoana catchment, B. Roma Valley/Maliele catchment, **8**. National boundary. Note that the national boundary in the west and southwest is along rivers.
Source: D.O.S. 621 (1:250,000, East & West Sheets, 1969).

terials such as maps, air-photographs, hydro-meteorological and geological information; vicinity to the National University of Lesotho; location within current, planned or recently completed land management project areas; and accessibility during all seasons.

Two catchments were chosen and within each of them smaller areas were selected for detailed studies. The two catchments are located in central and northern Lesotho and are centred around latitude 29° 25′S and longitude 27° 45′E, latitude 29°S and longitude 28°E respectively. The two drainage basins form part of the Caledon river basin which is one of the major drainage basins of the country (Fig. 1.1). The landforms within the two catchments are dominated by undulating and rolling dissected plains on Red Beds formation, sandstone escarpments and scree slopes, basalt mountain slopes with or without inflexions and gently undulating pediplains on basalt and sandstones above the 1 800 m contour (Bawden and Carroll, 1968). The vegetation is predominantly grassland of *Themeda-Festuca* and *Themeda-Cymbopogon-Eragrostis* transitional to Highland Sourveld.

The two catchment areas occupy two of Lesotho's major agricultural lands for crop and stock production. The lowlands also form the major settlement areas. The lowlands, areas below the major escarpment, have been classified to class 1 in agricultural potential. The rest of the areas are dominated by classes 4, 6 and 7 based on a 7-unit scale for suitability of crop production (Bawden and Carroll, 1968). The main problem in all these areas is a limitation caused by soil erosion in the form of gullies, rills and areas of bare ground without any top soil or vegetation and in some cases bare bedrock outcrops.

1.3 Soil erosion and reservoir sedimentation

The problem of soil erosion and sedimentation comprises detachment, transportation and deposition of sediments away from their original positions in the soil mass. The main agents involved are water, gravity, ice and wind. Soil erosion and sedimentation processes due to the agency of water form the central theme of the present studies.

Soil erosion is commonly classified as one of two types. The normal geological erosion which operates on the surface of the earth whenever there are energy flows in the form of water, wind or ice, and accelerated erosion caused by the activities of man which upset the balance between soil, vegetation cover and the erosive power of the various agents of geological erosion (Young, 1972; Hudson, 1971; Dunne, 1977; Faniran and Areola, 1978; Morgan, 1979). In the present studies the combined rates of the two types of erosion

are measured. However, it can be said that accelerated erosion is the dominant type because all the study areas are under intensive human use either as cultivated lands, pastures or settlements.

Soil erosion resulting from the action of water entails detachment of soil particles from the soil mass by raindrop impact and scouring by water flowing on or through the regolith, and transportation of the detached particles by splash and flowing water (Young, 1972; Hudson, 1971; Faniran and Areola, 1978; Bolline, 1978; Cooke and Doornkamp, 1978; Morgan, 1978). The result is two major forms of soil erosion, namely, raindrop (splash) erosion and runoff erosion. Runoff erosion can further be subdivided into two phases depending on whether the water flows on or through the regolith, namely, surface erosion (surface wash) resulting from overland flow, and subsurface erosion due to through- or interflow (Kirby, 1969; Young, 1972; Ward, 1975; Cooke and Doornkamp, 1978).

Surface erosion comprises four forms corresponding to progressive concentration of overland flow: sheet erosion, rill erosion, gully erosion and stream or channel erosion. These erosion forms correspond to unconcentrated overland flow, flow concentrated in micro-channels without any permanency, flow concentrated to well-established channels and flow in major streams and rivers (Hudson, 1971; Cooke and Doornkamp, 1978). In some works (Chow, 1964; FAO, 1965) raindrop erosion is included in sheet erosion. However, Hudson, using evidence from as early as the 1940s, argues that the term sheet erosion is misleading because it implies uniform removal of soil by an even flow of thin sheet of water. This implication is invalidated by the fact that runoff is seldom in thin sheets, and laminar flow scours only at velocities much higher than those usually occurring in nature (Hudson, 1971, p. 38).

Soil erosion by water is dependent on

the relationship between the erosivity of raindrops and running water and on the erodibility of soils (Bryan, 1968 & 1974; Hudson, 1971; Stocking, 1978; Stocking and Elwell, 1976; FAO, 1965; Morgan, 1979). The major variables in the erosion-sedimentation system are therefore climatic variables which can be summarized by rainfall indices of erosion, flow characteristics expressed by erosivity and soil variables represented by soil erodibility. The other controlling factors are vegetation cover, land use, topographic and surface properties like slope gradient, slope length, and surface roughness (Chow, 1964; Hudson, 1971; Cooke and Reeves, 1976; Cooke and Doornkamp, 1978; Faniran and Areola, 1978; Morgan, 1978). These variables have been combined in various formulas used in predicting soil loss from crop land and virgin lands. The most widely used of these formulas is the Universal Soil Loss Equation developed by Wischmeier et al. (1958, 1965, 1976). Good reviews of the history of development of these formulas, their usefulness and limitations are found in FAO (1965), Hudson (1971), Young (1976), and Arnoldus (1977).

Sedimentation forms can be put into two major groups depending on the distance between the point of detachment and deposition (transportation distance): (1) local sedimentation whereby the materials are transported relatively short distances before deposition in the form of colluvial and alluvial fans and floodplain deposits at the base of the hillslopes, (2) downstream sedimentation involving sediments which are transported across slopes to the main drainage channels and further transported by rivers and streams in the form of suspended-load, bedload and solution load. These are deposited where the flow velocities are reduced. The coarser fractions are deposited first. The amount of sediment transported in streams and rivers thus forms an important link in the soil erosion and

sedimentation system. The types of damage due to downstream sedimentation are: channel sedimentation, reservoir and lake sedimentation, and river plain sedimentation caused by overtopping of river banks and flooding the valley bottoms. The main sedimentation processes studied in the present studies are those connected with reservoir sedimentation and local deposition within the drainage channels, especially gully bottom fill deposition.

1.4   Previous studies of erosion and sedimentation in Lesotho

The problem of erosion and sedimentation in Lesotho has been dealt with previously in a broad view in various resource studies of Lesotho. The information up to 1970 was described by the author in a review of water and soil resources of Lesotho (Chakela, 1973). The major works in that review were the reports by Pim (1935) and Bawden and Carroll (1968). Pim's report was the first document to describe in detail the problems of soil erosion in Lesotho and relate the problem to the economic status of the country. Although no quantitative information was produced, erosion was declared, on qualitative basis, to be very severe. The report's conclusions and findings can be summarised as follows. The geological soil erosion in Lesotho is accelerated through overstocking, overpopulation, overcultivation, poor road construction and management methods, leading to gully erosion, channel erosion, sheet erosion, rill erosion and deterioration of pastures, and lack of well-planned soil conservation measures. The commission, on the basis of these observations, recommended: improvement of roads to minimize erosion caused by storm waters from the road drains and to limit the number of footpaths to areas where erosion was minimal or could be controlled,

12

initiation of extended scheme of dealing with soil erosion spread over a period of 10 years, including installation of simple anti-erosion measures throughout the country, ecological survey of the country, and construction of small irrigation works.

The Bawden and Carroll (1968) study is a follow-up of the 1960 Economic Survey Mission (Morse, 1960). Concerning the problem of soil erosion, their study reached the same conclusions as those reached by Pim in 1935 despite the fact that several projects and schemes were planned and implemented in the intervening period. The causes of the severe soil erosion situation in Lesotho were associated with overstocking, improper land use methods and lack of understanding among the indigeneous population of the necessity for devising and maintaining soil-conservation measures, and following the recommendations of the surveys. Therefore a recommendation was made for introduction of more strict conservation policy and education of the local inhabitants.

There was then no quantitative data on the rates of erosion and sedimentation in the country. The 1970s saw a great change in collection of data on erosion and sedimentation, although these data were still marginal to the general evaluation of water and soil resources development studies (Binnie and Partners, 1972; West, 1972; Carroll et al. 1976; Jacobi, 1977; Flannery, 1977). Relevant to the investigations in the present studies are the information on discharge and sediment transport of the major drainage basins of Lesotho (Binnie and Partners, 1972; Jacobi, 1977), soil surveys within project areas (West, 1972; Carroll, et al., 1976), and gully erosion studies (Flannery, 1977). The estimated sediment yield of the Caledon river basin was found to range from 1400 m^3 km^{-2} y^{-1} to 2000 m^3 km^{-2} y^{-1} (1.2—1.7% of mean annual runoff) (Binnie and Partners, 1972). The estimated dry bulk density of the sediments was taken as 1 t m^{-3} yielding a sediment load value of 1400—2000 t km^{-2} y^{-1} for the Caledon basin.

In reporting of streamflow data, Binnie and Partners (1972) divided the country into four major drainage basins: Orange River Drainage Basin, 90.7 m^3 s^{-1} mean annual discharge, Maphutseng Drainage Basin, 1.2 m^3 s^{-1} mean annual discharge, Makhaleng Drainage Basin, 14.2 m^3 s^{-1} mean annual discharge, Caledon Drainage Basin, 24.1 m^3 s^{-1} mean annual discharge giving a total mean discharge for the whole country to be 130.2 m^3 s^{-1}, corresponding to mean annual runoff of 137 mm.

Information pertinent to the areas studied in the present investigation are the data on the Caledon river basin. Therefore it is important to look closely at the data obtained in the tributary basins to the Caledon. The tributaries of the Caledon river drainage basin studied by Binnie and Partners (1972) and Jacobi (1977), relevant to the present studies are:

1. Hlotse river above Khanyane (728 km^2).

2. Phuthiatsana river above Mapoteng (386 km^2)

3. Little Caledon above Masianokeng (945 km^2)

The first two catchments have characteristics similar to the Khomo-khoana catchment and are very close to it, the third catchment has Roma Valley and Maliele catchments as its subcatchments.

The estimated mean annual discharge and runoff for Hlotse river at Khanyane and Phuthiatsana river at Mapoteng are 5.0 m^3/s (216 mm/year) and 2.2 m^3/s (181 mm/year) respectively. For the Little Caledon river at Masianokeng a mean annual discharge of 3.1 m^3/s (103 mm/year) was obtained as the best estimate (Binnie and Partners, 1972). Jacobi (1977) reported sediment yield for the same station to be 1979 t km^{-2} y^{-1} based on 6 years of water discharge measurement and 250 water samples. He gives mean concentration of dissolved solids to be 180 mg/l. The dissolved load corresponds to 2% of annual sediment load. For the Phuthiatsana river at Mapoteng, Jacobi gives the sediment yield to be 2968 t km^{-2} y^{-1}.

The conclusion that can be drawn from these studies are that the best estimates for mean annual runoff for the areas studied in the present project range from 100—200 mm/year with sediment yields varying from 200 t km^{-2} y^{-1} to 3000 t km^{-2} y^{-1} with the highest values in the lowland catchments.

In their conclusion to the estimation of sediment yield from the various drainage basins in Lesotho, Binnie and Partners (1972) pointed out the following factors which affect distribution of sediment yield in Lesotho drainage basins and should be taken into consideration when estimating sediment yield from the smaller catchments within the Caledon river drainage basin:

1. Lowland rocks and soils derived from them are easily eroded compared to those of the highlands.

2. Rainfall – sheet erosion is highly dependent on the erosive power of the rainfall and it is possible that higher rainfall areas have more intense rainfalls.

3. Vegetation—good vegetation cover reduces the erosive power of rain, therefore the degree of vegetation deterioration (overgrazing, urbanization and cultivation) influences the rate of erosion by rain.

4. Gully erosion—gullies in certain lowland catchments make up to 60% of the erosion-affected areas.

5. Cultivation—cultivated areas are more sensitive to erosion by rainstorms and wind than areas under natural vegetation cover.

Jacobi (1977) pointed out the high variability of sediment yield on a regional basis depending on bedrock, basin slope and extent of cultivation and that the Little Caledon on average transports 25% of total year's sediment in 2 to 4 days.

The soil survey reports supply very little quantitative data, but have improved the knowledge concerning soils and their relative erodibility. The erodibility of the soils within the study areas (Khomo-khoana and Thaba-Bosiu areas) have been estimated and an attempt has been made to estimate the rainfall factor for the whole country (Carroll et al., 1976; Flannery, 1977). However, the studies give very little information on the rates of different erosion processes. The maps produced in connection with soil surveys provide good starting points for the inventory of gully erosion, sheet erosion on non-cultivated land, and sedimentation along the main streams. Rill erosion and sheet erosion on cultivated lands is not clearly revealed on the air-photo mosaics used for soil surveys, but combination of these maps with colour photography could be used to supplement and extend the field studies to 1971.

Some of the results obtained in the present studies have been reported in three earlier publications by the author (Chakela, 1974, 1975 and 1980). The first of these reports dealt mainly with the gully types, location and gully growth mechanism, because gully erosion is the most spectacular erosion form in the studied areas. The gullies in the study areas can be placed into three classes on the basis of location within the catchment (Brice, 1966): valley-bottom gullies, valley-head gullies, and valley-side gullies. They normally form a dendritic pattern similar to that of the drainage system. Three modes of gully growth mechanism are dominant in the studied areas: side-gully development resulting from a system of cracks on the main gully side due to expansion and contraction of clays on wetting and drying and due to root pressure, side-gully formation and headward advance of the gully-head scarp through piping, and lateral and headward growth of the major gullies through undermining of the side walls by flowing water in the channel or splash at the gully-head scarps. Piping was found to occur either at the bedrock-overburden interface or above a crustal layer (claypan or concretional layer).

The second report (Chakela, 1975) is a summary of the observations and measurements made in the field period 1974/75 rainy season. The main observations and conclusions reached after the first year of field study and observations can be summarized as follows. Reservoir sedimentation was found to be of the order of 15 cm to 100 cm per year, sediment concentration in the Roma Valley catchments showed apparent increase with increasing distance from the head waters of the Roma Valley, the annual rate of gully-head advance ranged from 50 cm to over 10 metres, the existing counter-erosion measures have very little effect on local sheet and rill erosion on the cultivated lands and in some cases these measures have aggravated gully erosion.

The third report (Chakela, 1980) deals with reservoir sedimentation within Roma Valley and Maliele catchments. The rate of reservoir sedimentation was found to vary from 0 to 20 cm/year or sediment accumulation of up to 4000 tonnes during the study period, with mean annual rates of up to 1000 tonnes within the reservoirs studied.

2. Regional setting of the study areas

2.1 Introduction

The purpose of this chapter is to give background information on the physical properties of the drainage basins within which the study areas are located. The description varies in detail depending on the availability of data and the amount of previous information in the catchments concerning the properties described. The main subjects dealt with are grouped under five headings: bedrock geology; climate and hydrology; landforms and surficial deposits; soils; vegetation and land use. In the description of each of these physical properties reference is made to the effect of the properties on the processes under study.

2.2 Roma Valley and Maliele catchments

The study area consists of two catchments located approximately 29° 25' to 29° 30' S and 27° 40' to 27° 50' E in central Lesotho within the Maseru district (Figs. 1.1 and 2.1). The catchments are 5th and 4th order catchments with relief ranging from 1585 m to just over 2500 m above m.s.l. for the Roma Valley and to 1945 m for the Maliele catchment. The two catchments form some of the uppermost subcatchments of the Little Caledon river catchment which in turn is a subcatchment of the Caledon drainage basin. Both catchments lie within the Thaba-Bosiu Rural Development area which was initiated in 1972 and completed in 1979. The Roma Valley area below the escarpment (1800 m) has been a project area for the Roma Valley Agricultural Project since 1968.

2.2.1 Bedrock Geology

The bedrock geology of the Roma Valley and Maliele catchments consists of the following five formations:

Drakensberg Beds and Dolerite Dykes

The mountainous part of the catchments and the foothills region are underlain by a series of basaltic lava flows. The various outpours form moderately strong, grey to dark grey layers of thicknesses varying from a few decimetres to possibly over 30 metres (Binnie and Partners, 1972). The dominant minerals in these lavas are plagioglase feldspars, pyroxenes and some olivines (Cox and Hornung, 1966). There are some clay minerals in the ground mass consisting mainly of montmorillonite and vermiculite. At about 1800 m to 1900 m the flows come into contact with the underlying sandstone formations. Intruded into the basalt and the older formations are dolerite dykes which are said to be contemporaneous with the basalt flows (Stockley, 1947). These dykes have formed ridges criss-crossing the foothills and mountain regions. Seven of these ridges form very prominent lineations within the catchments and form passes from the lowlands to the highlands.

Cave Sandstone Formation

The Drakensberg Beds are underlain by Cave Sandstone formation which consists of buff-coloured, weakly to moderately cemented, fine-grained sandstone. Within the Roma Valley and Maliele catchments the formation is exposed in lower parts of the foothills region and consists of a cliff line with caves, broken through here and

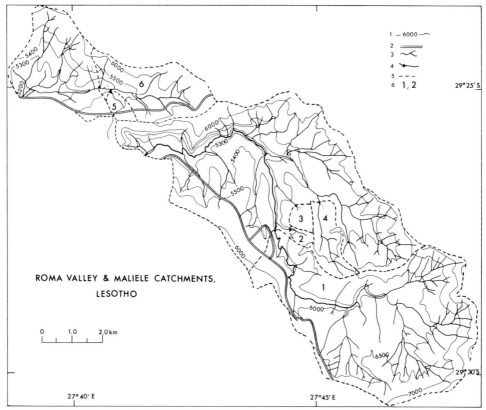

Fig. 2.1. Map of Roma Valley and Maliele catchments and the location of study areas. **1**. Contour lines in feet with 500 feet interval, **2**. Roads, **3**. Hydrography, **4**. Investigated reservoirs, **5**. Catchment boundaries, **6**. Investigated subcatchments.
Source: D.O.S. 421 (1:50,000, 1955—66).

there by dolerite dyke depressions or gorges formed by the main streams. The mineral composition of the formation is mainly subangular quartz and feldspar grains with calcareous cement (Stockley, 1947; Binnie and Partners, 1971). The formation is poorly bedded and has inclusions of clay-shales and silty inlayers at the base. The origin of the formation is largely attributed to aeolian processes resulting in uniform grain size and lack of bedding planes (Binnie and Partners, 1971).

Transition Beds Formation

The Cave Sandstone escarpment within these two catchments forms the uppermost sandstone area. Below this escarpment are alternating layers of sandstones, mudstones and clay-shales. In some reports (Binnie and Patners, 1972), these are grouped together with the rest of the underlying formations as shales, mudstones and sandstones and the topmost of these are included in the Cave Sandstone formation, but Stockley following Van Eeden, separates them from the massive sandstone above and calls them the transition beds (Stockley, 1947, p. 40). They consist of red and buff sandstone layers alternating with purple and red to blue shales and mudstone layers, giving the scree slopes a terraced form. The terrace-like form of the slopes is attributed to differences in weathering between the sandstone layers and those of mudstones and clay-shales. This formation is transitional in two meanings: the grain size composition of the sandstone layers is very similar to that of the Cave Sandstone formation while the inclusions of the red coloured mudstones are more similar to the underlying Red Beds formation.

Red Beds Formation

The rest of the catchment, except in the lowest parts of the area, close to the main stream channel, is underlain by sandstones, shales, and mudstones of the Red Beds formation. In comparison with the Transition Beds, the Red Beds sandstones are coarser and thicker. They also contain silicified wood. The formation has coarse grits at the base and thus the transition to the underlying formation is very diffuse.

Molteno Beds Formation

The lowest parts of the catchment are in the area of Molteno Beds consisting of white arkosic grits and sandstones with blue, gray or greenish shales. These sandstones form the lowest spurs in the Little Caledon basin and the larger part of them is not exposed within the Roma Valley and Maliele Catchments.

Recent Formations

Superimposed on these older formations are recent, semi-consolidated to loose deposits consisting of alluvial deposits along the main drainage channels, colluvium on the scree slopes and windblown sands on the open spurs of the rolling and undulating lowland terrain. To these recent formations should be included weathering products of basalt and sandstone in the basalt and sandstone terrain respectively and the deeply weathered materials within the dolerite dykes.

The recent deposits are now being entrenched by a system of gullies and streams to varying degrees of incision. The entrenchment of these deposits is dealt with in Chapter 4. Table 2.1 gives the summary of the formations described

Table 2.1. *Geological Succession and Formations in Roma Valley, Maliele and Khomo-khoana Catchments (After Stockley, 1947, Binnie and Partners, 1972, and Dempster and Richard, 1973).*

Age	Series	Formation (Maximum thickness)	Description
Tertiary to Recent			Pedisedimentary and aeolian gravels, sands, silts, clays and weathering products.
Early Jurassic			Instrusive dykes of basalt, dolerite and gabbro
Early Jurassic	Stormberg	Drakensberg Beds (500 m)	Hard, dense lava flows consisting of basalt. Ashy or agglomerate beds near the base in some places.
		Cave Sandstone (240 m)	Buff coloured, fine-grained sandstones (very occasionally bedded) with occasional clay-lenses or mudstone lenses.
		Transition Beds (80 m)	Red, fine-grained sandstone with some clay-shales and mudstones.
Triassic		Red Beds (260 m)	Buff, red sandstone alternating with thin shales and mudstones.
		Molteno Beds (150 m)	White, coarse-grained sandstone with grits and gray or greenish clay-shales and mudstones.

above as mapped by Stockley (1947). The table contains only those formations exposed within the study areas.

2.2.2 Climate and Hydrology

The climate of the Roma Valley and Maliele catchments can be described with the help of the limited data that are provided by the rainfall stations at the National University of Lesotho (N.U.L.) and at Christ the King High School (C.K.H.S.). The rainfall records for the two stations start in 1935 and 1948 respectively. However, like many stations in Lesotho, several gaps exist in the record. Discharge and runoff data are completely lacking in the area and the only data relevant to stream flow regime in the catchment are those done in connection with the present studies.

In general the climate of the area can be described as temperate, continental, sub-humid with mean annual rainfall ranging from 810 mm in the lowlands to over 1000 mm in the mountain province. The larger portion of this rainfall occurs during the rainy season of October to April in the form of high intensity thunderstorms. The wettest month is normally January but in some years the shift may be to February and March. The driest months are June and July. The winters are cold and dry and have several sub-zero temperatures at night. Mid-day in winter is normally warm and sunny. The windiest months are August and September with light to moderate duststorms.

Temperature

Temperature data for the two stations within the Roma Valley are hard to obtain but two sets of information are published for the N.U.L. station (Binnie and Partners, 1972; Ministry of Works, 1970). Since April 1976 data have been published from the two stations in the Climatological Bulletin of the Meteorological and Hydrological Services. Fig. 2.2. summarises the mean monthly temperatures at N.U.L. based on data derived from these sources. It should be borne in mind that these figures cover only a period of, at the most, five years of complete record.

Precipitation

Although the rainfall record covers only a 40-year period, it is relatively good with very few gaps. The record for the period 1936—1977 is unbroken and gives a mean annual rainfall of 824 mm (Fig. 2.3). For the analysis of the variation in the annual rainfall, a 30-year period, 1948—1977, was used. The period shows large variations in annual rainfall with four dry periods. The dry period which started in 1964 and ended in 1973 is interrupted by only one year with annual rainfall above the mean for the period and it contains the driest year for the 30-year record (Figs. 2.3—4). The earlier dry period in the available record is that of 1945—1948. These dry periods are significant for the vegetation cover in the area and its ability to abate soil erosion and for water supply for agricultural and domestic purposes. The oral information relates fresh entrenchment of gullies and disappearance of swamps and reed meadows to these periods, with the highest gully activity concentrated to the high annual rainfall years following immediately after the dry years. The small lake which used to exist within Maliele catchment is reported to have dried up

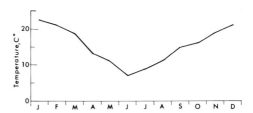

Fig. 2.2. Mean monthly temperature at Roma, 1969—1976.
Source: Binnie and Partners, 1972; Climatological Bulletin, 1974—77.

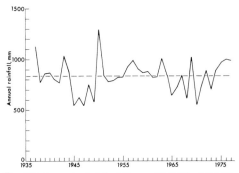

Fig. 2.3. Annual rainfall at Roma, 1936/37–1976/77. Water years.
Source: Meteorological data to Sept. 1970. Climatological Bulletin, 1974–1977.

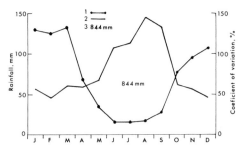

Fig. 2.4. Mean monthly rainfall and the coefficient of variation, 1948–77, at Roma. **1**. Mean monthly rainfall, **2**. Coefficient of variation, **3**. Mean annual rainfall for the period.
Source: Meteorological data to Sept. 1970. Climatological Bulletin, 1974–1977.

finally during the 1965 drought and since then even the reed meadow associated with the lake has disappeared.

Annual distribution of rainfall for the 30-year period is shown in Fig. 2.4 together with the coefficient of variation per month expressed as the ratio of standard deviation to the mean value. The dry season has low monthly means but the variation is highest during this period.

During the study period, which started immediately after the 1970 drought, it was observed that the rainfall monthly totals were successively increasing from 1973 and started to drop again in 1976 (Fig. 2.5). During the period, two extreme values of rainfall were observed and both coincided with approximately 45-minute thunderstorms leading to very high, short duration stream flows. Annual rainfall shows an increase from the beginning of the study period from 608 mm in 1970 to a maximum of 1093 mm in 1976. Comparison with the 30-year mean shows that in the study period the rainfall in the area was relatively higher than normal except for 1973 where rainfall was 70 mm less than normal for the station.

The rainfall station at N.U.L. does not, however, represent all the rain that contributes to runoff and streamflow of the catchment. Several heavy rainstorms have been observed during the rainy seasons 1973–1977 in the upper reaches of the

catchment with no rainfall in the Roma Valley lowlands. Therefore, to get a complete picture of the relationship between rainfall and runoff and discharge over the catchment, several stations are needed in the foothills and mountain zones of the catchment.

Streamflow

No stream gauging has been made in this area. However, the two catchments form the highest reaches of the Little Caledon catchment, which has a record of stage and water discharge observations extending as far back as 1965. Runoff estimates have been made for the Little Caledon catchment using observations at Masianokeng. A figure of 103 mm for the mean annual runoff was obtained by Binnie and Partners using 5-year data from 1965–1970 (Binnie and Partners, 1972). This estimate applies to the whole of the catchment area of 945 km² compared to the combined area of Roma Valley and Maliele catchments which is only 80 km². Therefore the runoff of the catchment can be said to lie somewhere around 100 mm per year or higher for these catchments, if one takes into account that the largest part of the Little Caledon catchment is in the rainfall-poor lowlands, while the catchments studied in this investigation have

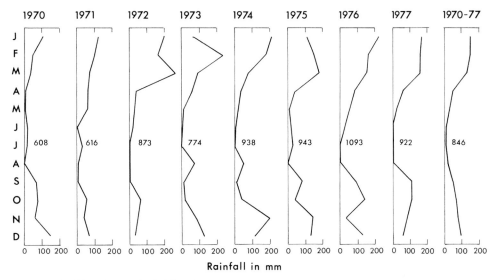

Fig. 2.5. Monthly and annual rainfall at Roma, 1970—1977.
Source: Meteorological data to Sept. 1970. Climatological Bulletin,1974—1977.

most of their runoff resulting from rainfall-rich, steep mountain slopes.

Mean annual runoff can be estimated also by the use of the modified Langbein method (WMO, 1970). The method uses mean annual precipitation, mean annual temperature in cm and °C respectively, and monthly values for these two parameters. The same figure (100 mm) was obtained using the N.U.L. data as that obtained in the study by Binnie and Partners. If, however, the months with precipitation greater than 50 mm are used, the estimated mean annual runoff becomes

85 mm or 10.5% of mean annual precipitation.

The two catchment areas are drained by streams of various types. The main streams in each catchment are perennial but the winter flows are so low that they border on intermittent. The major tributaries of the main streams are ephemeral and intermittent creeks, the majority of which are deeply encised gullies fed by springs oozing at the gully-heads. The flow in these gullies is limited to the rainy season and for the majority of them to periods immediately after or during rainstorms.

Table 2.4. *Rainfall Frequency at Roma, 1973—1977.*

Rainfall intensity	Frequency, days					
mm/day	1973	1974	1975	1976	1977	1973—1977
1—5	52	63	43	48	48	254
5—10	14	20	1	21	20	85
10—20	19	20	12	23	22	96
20—30	6	10	11	5	6	38
30—40	1	1	2	5	0	9
40—50	0	0	1	2	1	4
50—60	0	0	0	0	2	2
60—70	1	0	0	0	1	2
	93	114	79	104	100	490

For typographical reason, Table 2.4 has been placed here to coincide with the text.

21

These flows are characterized by flashy floods of very short duration and may be over in a period of 10—15 minutes. These characteristics are dealt with in the appropriate sections of Chapter 4 of this report.

In order to make up for the lack of hydrological observations on streamflow, gauging stations were started in 1974 at the outlet to the catchments and some observations of water stage and discharge were made. The results of these limited observations are dealt with in Chapter 4 of this report under the headings Water and Sediment Discharge. The record obtained covers only the rainy seasons of 1975/76, 1976/77 and part of 1974/75. The following observations can be noted: high water flows occur in the form of flash-floods with short-lived duration varying from 15 minutes and seldom over 45 minutes depending on the types, intensities and duration of the rainfall producing the flows. During heavy thunderstorms, most of the ephemeral streams (dry gullies) exhibit flows with a special flow form; two of which were observed during the present studies. The flows are 5—30 cm high in a broad front, 5—20 m wide, tapering up-stream where the height diminishes to zero or a very thin layer. This was observed after the first rainstorm after a long spell without rain. The front is mostly heavily loaded with plant debris and sediment. The water downstream of the front is either clear or the ground is made up of mudcracks.

2.2.3 Landforms and Surficial Deposits

The landforms and surficial deposits within the Roma Valley and Maliele catchments have been described generally in the works of Bawden and Carroll (1968), Binnie and Partners (1972), Chakela (1974) and P. H. Carrol et al. (1976). The last two reports deal specifically with the studied areas while in the other works the landforms within the study areas are dealt with as part of the general geomorphology of Lesotho.

In the present context Bawden and Carroll's (1968) Land System divisions (Table 2.2) have been used as the main landform groupings for the description of the landforms and loose deposits within the catchments. The published information

Table 2.2 *Land Provinces, Land Regions, Agro-ecological Zones and Land Systems: Roma Valley, Maliele and Khomo-khoana Catchments (After Bawden and Carroll, 1968).*

Land Province	Land Region	Agro-ecological Zone	Land System (altitude, m)	Occurrence
Mountain	Lower Mountain slopes	Lower Mountain Grazing	North-west Escarpment (1950—2300)	Khomo-khoana Catchment
			Compound Lower Slopes (2200—2500)	Roma Valley Catchment
	Foothills	Foothills	Northern Basaltic Foothills (1800—2200)	Roma Valley, Maliele and Khomo-khoana Catchments
			Southern Basaltic Foothills (1800—2200)	Roma Valley/Maliele Catchment
Lowland	Lowlands	Lowland	Lowlands Escarpment (1650—1900)	Roma Valley, Maliele and Khomo-khoana Catchments
			Central Lowlands (1560—1700)	Roma Valley/Maliele Catchments
			Red Beds Plains	Khomo-khoana Catchment

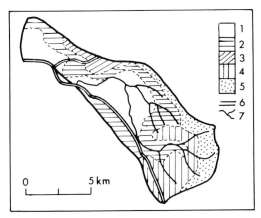

Fig. 2.6. Roma Valley and Maliele catchments: Land systems. **1**. Central Lowlands, **2**. Lowlands Escarpment, **3**. Northern Basaltic Foothills, **4**. Southern Basaltic Foothills, **5**. Compound Lower slopes, **6**. Roads, **7**. Hydrography.
Source: Bawden and Carroll, 1968.

has been supplemented by the use of airphoto interpretation, analysis of topographic maps (1:50 000 and 1:100 000) and field observations in order to present a more detailed description of the geomorphology of the area. This description starts at the head waters of the Roma Valley catchment and proceeds to the outlet of the two catchments. Fig. 2.6 is a summary of the land systems in the area as mapped by Bawden and Carroll.

Compound Lower Slopes Land System

This land system occupies the highest reaches of the Roma Valley catchment (Figs. 2.7—8) and consists of mountain crests, at elevations varying from 2300 to just over 2500 m above m.s.l.; steep, straight to slightly concave mountain slopes, forming a semi-circular valley-head slope. The slopes are dissected by several minor streams producing crested to rounded interfluves and minor V-shaped valleys. The slopes are covered mainly with grasses but shrubs dominate along the drainage channels and on most of the south to south-west exposed slopes. The

loose deposits in this land system consist mainly of basalt weathering products, dolerite weathering products, alluvium and colluvium along the major drainage lines, and from the slopes.

The dominant processes are linear erosion by streams; surface wash on the side slopes and interfluves; very limited rilling and almost no gully formation. The gullies are limited to areas near the drainage channel and entrench the small pockets of alluvial deposits in these zones.

The area is heavily overgrazed with short grass stands, which still form a complete carpet cover of the ground except on the steepest slopes where areas of exposed bedrock form areas of initiation of sheet flow after rainfall. Another significant erosion feature in this land system is the formation of furrows along livestock routes. In areas where paths run down the slope, rilling has been initiated and the furrows deepened to over 30 cm or to the bedrock.

Basaltic Foothills Land System

In the Land Resource study of Lesotho by Bawden and Carroll (1968), this land system is divided into two land systems named Northern and Southern Basaltic Foothills. Within the Roma Valley the difference between these two is so minor that here they are treated as one. The land system consists of two planation surfaces forming gently undulating pediplains separated by a minor basalt scarp. The elevation within the land system is from 1820 m to just over 2200 m. The upper planation surface is entirely on basalt and consists of isolated benches, the largest of which is in the southernmost part of the upper reaches of Roma Valley catchment (Figs. 2.7 & 2.8) with the central parts consisting of minor residual benches at the foot of the mountain slopes. The minor escarpment below this upper surface is either steep (Fig. 2.9) or consists of gently sloping convex spur slopes separated by small streams with minor terraces formed

23

Fig. 2.7. Mountain slopes (background) and Basaltic Foothills, Roma Valley area 1 (Photo: Q. K. Chakela, April 1977).

by different layers of lava beds.

The lower planation surface is a structural platform of Cave Sandstone and is almost level in places or forms broad level to rounded spurs with topographic depressions formed in the bedrock. The major streams have cut a deep, branched gorge into the surface (Fig. 2.10).

The vegetation is mainly grasses and some shrub groves along drainage channels and minor valleys descending the minor scarp separating the two surfaces, especially on south to south-west facing slopes.

The loose deposits consist of colluvial, alluvial deposits and some aeolian sands on the open broad spurs of the sandstone plateau. The materials are all very fine and very few gravels are found in the drainage channels where the pebbles may dominate as most of the finer material is washed off by rivers. The water is always

Fig. 2.8. Basaltic foothills and mountain slopes landform zones, Roma Valley area 1. The level surface in the foreground is the residual planation surface on basalt at about 2200 m. Terraces on the mountain slopes mark the different lava flow layers. (Photo: Q. K. Chakela, Jan. 1975.)

24

Fig. 2.9. A small escarpment separating the Cave Sandstone plateau from the basaltic foothills planation surface. Several slide scars and small earthflows were observed on these slopes. The largest of the earthflows can be seen in the centre of the picture. (Photo: Q. K. Chakela, Jan. 1975.)

Fig. 2.10. Cave Sandstone plateau with the Maphotong gorge in the centre and basaltic foothills tops in the background. (Photo: Q. K. Chakela, Jan. 1975.)

Fig. 2.11. Regolith stripping above the escarpment, and rockfalls and cave formations below the escarpment, Roma Valley area 1. (Photo: Q. K. Chakela, Jan. 1975.)

very clear, even during rains.

The dominant processes in this land system, starting from the lower parts to the highest, are regolith stripping along the lower edges of the sandstone plateau (Fig. 2.11); bank erosion and limited side gullying and deposition within the major streams; sheet wash and rilling on most of the cultivated areas on the broad spurs of the Cave Sandstone plateau and at the transition to the upper steep slope of the minor escarpment separating the two planation surfaces. Minor land slides and earth flows have been observed in some places on the steep escarpment slopes (Fig. 2.9). Very few gullies exist in this land system, which may be due to shallowness of the soils, the clayey characteristics of the basalt-derived soils or the better vegetation cover found in the depressional zones. Side gullies are formed at two locations in the area:

a. Side gullies to the main streams extending 10—50 m and very broad (compared to length) and terminating on bedrock scarps in the upper parts (Fig. 2.12).

b. Side gullies along footpaths crossing the streams and deepened rills at the upper parts of the sandstone plateau mainly on cultivated fields.

Another process worth noting in this land system is the supposed deep weathering of sandstone producing shallow topographic depressions filled with loose deposits of almost unimodal grain size distribution. Several of these depressions have no surface channels but are either

Fig. 2.12. Spur-side discontinuous gully starting at a deeply entrenched footpath and ending on the alluvial plain of a small stream channel formed in the bedrock depression within the Cave Sandstone landform zone. (Photo: Q. K. Chakela, Jan. 1975.)

bogs with springs at the lower parts or are used as water-wells by the villagers.

Lowlands Escarpment Land System

The Lowlands Escarpment land system consists of two distinct landforms (Figs. 2.13 & 2.14).

1. The Cave Sandstone scarp which forms a cliff or bluff line separating the lowlands from the uplands. It consists mainly of a vertical cliff, steep, stepped slopes or convex downward slopes. In the upper Roma Valley area this includes a gorge cut by the main stream to the depth of approximately 200 m. Where the dolerite dykes cross it depressions have been formed and these are used as passes to climb into the uplands.

2. Straight scree/colluvial slopes with debris of pebbles and boulders embedded in fine colluvial materials. These straight slopes are separated by pockets of depressions with thick, stratified colluvium at the heads of most of minor lowland streams. Most of these deposits are heavily gullied to the bedrock. The lower parts of the scree slopes either terminate on pediments, minor alluvial fans or continue straight into the alluvial plane of the major streams.

Vegetation is mainly grassland but all the south to south-west facing scree slopes are clad with some shrub vegetation. Within the Roma Valley a semi-forest plantation of wattle trees covers the northeast facing slopes above the Roma Mission. The main processes in this landsystem are gullying, minor landsliding and infrequent rockfalls resulting in rocky screes below the escarpments.

Central Lowlands Land System

The Central Lowlands land system can be divided into three landfacets (see also illustrations in 4.1):

1. Pediment and spur slopes
2. Valley flats
3. Lowlands topographic depressions

The valley flats form the lowest parts of the catchment along the main drainage channels. They are narrow zones with thick alluvial deposits, dark brown to dark grey and black in colour and silty to silty loam and clay loams or clays in texture. The lowest reaches of the Roma Valley

27

Fig. 2.13. Cave Sandstone Escarpment and scree slopes, Roma Valley above Maphotong. (Photo: Q. K. Chakela, April 1977.)

catchment have developed some minor terraces and minor meanders, but not as well defined as in the case of the valley flats zone in the Red Bed Plains land system.

Deposition within this landfacet occurs only in the form of point bars but not as mid-main channel accumulations as seen in the case of the Khomo-khoana river (Section 2.3.2).

Three pediment slopes can be distinguished within the Roma Valley area, but these have no abrupt break with the inter-tributary spurs found in Khomo-khoana catchment pediments. The inter-spur depressions are heavily gullied. Gullies are either actively eroding and extending headwards, or secondary scarps have been formed in the floors of the gullies and these are actively consuming the gully-fill deposits within the old gully walls or have been partially stabilized through construction of reservoirs, or by the fact that

equilibrium has been reached between gully width and depth.

Fig. 2.14 illustrates the essential features of the landforms within the Roma Valley catchment. These are discussed in more detail in connection with erosion and sedimentation studies in smaller catchments (4.1—4.6).

2.2.4 Soils

The major works, in which the soils of Roma Valley and Maliele catchments are described, are: Carroll and Bascomb (1967), Binnie and Partners (1972) and P. H. Carroll et al. (1976). This information was later summarized by Chakela (1974) (see also Schmitz, 1980). The works of Carroll and Bascomb, and Binnie and Partners cover the whole of Lesotho and Lesotho lowlands respectively. The study areas in the present report are part of the

28

Fig. 2.14. View of the Roma Valley area upstream of the National University of Lesotho. (Photo: Q. K. Chakela, Feb. 1976.)

area studied by P. H. Carroll et al.

In the summary report by P. H. Carroll et al. (1976), a soils classification table is produced which is given here as Table 2.3. This table is of interest for the general comparison of Lesotho soils with international soil maps. For this study the point of interest lies in the first column of the table dealing with soil series. These soil series form topo-sequences, closely related to bedrock geology, exposure of the slopes, elevation and landforms (Fig. 4.3). In order to enable comparison with the erosion studies, these topo-sequences are described below and exemplified by their major members. The member (series) given first is the highest on the mountain or hill slope in the mountain and foothills region, and

Table 2.3. *Soils within Roma Valley and Maliele Catchments as mapped by P. H. Carroll et al., (1976).*

Soil Series	Family	Subgroup	Group	Suborder	Order
Matšaba	Fine-loamy, mixed, mesic	Typic Argiudolls	Argiudolls	Udolls	Mollisols
Ralebese	Fine-loamy, mixed, mesic	Lithic Hapludolls	Hapludolls	Udolls	Mollisols
Maliele	Fine-loamy, mixed, mesic	Cummulic Hapludolls	Hapludolls	Udolls	Mollisols
Leribe	Fine-loamy, mixed, mesic	Typic Hapludolls	Hapludolls	Udolls	Mollisols
Khabo, thin	Fine-loamy, mixed, mesic	Typic Argiudolls	Argiudolls	Udolls	Mollisols
Khabo	Fine-loamy, mixed, mesic	Typic Argiudolls	Argiudolls	Udolls	Mollisols
Maseru, dark	Fine-loamy, mixed, mesic	Cummulic Haplaquolls	Haplaquolls	Aquolls	Mollisols
Maseru	Fine, mixed, mesic	Typic Albaqualfs	Albaqualfs	Aqualfs	Alfisols
Tsiki	Fine-loamy, mixed, mesic	Typic Albaqualfs	Albaqualfs	Aqualfs	Alfisols
Phechela	Fine, montmorillonitic mesic	Typic Pelluderts	Pelluderts	Uderts	Vertisols

the highest or northernmost for the scree slopes and lowland areas. In each sequence, thickness, drainage, texture and organic content tend to change downslope with most central members having impeded drainage, heavy texture and high organic matter. This is revealed most clearly in the foothills and lowlands where the colour changes from light to dark as one goes down slope, with hillcrests and spur crests covered by reddish to pale-yellowish soils of coarse texture (sands) and the depressions by dark-coloured clay soils.

Mountain Slopes Topo-sequence

The soils in this sequence are dominated by three soil series, dark-coloured, favourable structure, slightly to moderately acid grassland soils. The typical sequence occupies mountain slopes with straight upper portions and concave bases ending on a basalt planation surface or residual hill of the degraded surface.

Members of the sequence are:

1. Basalt Rockland: Bedrock outcrops with thin, stony soils and very sparse vegetation consisting mainly of grasses and herbs
2. Popa loam or fine sandy loam on basalt saprolite
3. Matšana silty loam to clay loam, normally a central member
4. Fusi silty loam to clay loam
5. Ralebese loam or fine sandy loam, end-member on the upslope side of a hillcrest.

The thickness of the solum varies from zero for the rocklands to over 100 cm for the central members but not over 120 cm. The general elevation of the sequence is 2000—2500 m above m.s.l.

Foothills Topo-sequence (planation surfaces and basalt spurs)

The topo-sequence represents soils found at elevations between 1830 m and 2000 m or on the minor escarpment between the Cave Sandstone plateau and the 2000 m planation surface within upper Roma Valley. Four typical members of this sequence are dark-surfaced, reddish, loamy to clayey in texture and generally on the top of thick deposits or saprolite. The sequence is typically represented by:

1. Ralebese loam to fine sandy loam
2. Matšaba loamy soil
3. Machachce clayey soils, forming along the lower, and more gently sloping parts of the slope or concave break/change of slope.

Foothills Depression Topo-sequence.

These are the soil series that flank the drainage depressions in the foothills and occupy some of the depressions in basalt terrain. They occur at about the same elevations as the Foothills Topo-sequence. They are normally represented by the following series in a north-south alignment:

1. Thabana clay loam, silt loam or loam
2. Phechela silty clay loam, silty clay or clay, central member along the drainage depression centre
3. Matela fine sandy loam, mostly on north to north-east facing slopes.

Cave Sandstone Plateau Topo-sequence

Unlike the topo-sequences described above, this topo-sequence is formed entirely on sandstone bedrock. It consists of five members, if the escarpment rockland is included, with the central member occupying the centres of drainage depressions or any other topographic depressions.

1. Ntsi fine sandy loam on sandstone hillcrest slopes and side slopes
2. Berea loamy fine sand
3. Thoteng fine sandy loam or loamy fine sand, on footslopes flanking the drainage depressions
4. Theko fine sand or fine sandy loam

5. Sandstone rockland—Cave Sandstone outcrops at the lower edge of the sandstone plateau including the escarpment cliff.

Colluvial/Scree Slope Topo-sequence

This is the topo-sequence with very few members and is dominantly colluvial deposits with large lobes strewn with boulders. Where the topo-sequence is complete it consists of the following two members:

1. Colluvium with well developed stratifications and boulders buried in the deposits. The content of fine materials and boulders differ from slope to slope. Most of the side slopes are covered with coarse grained deposits with very limited fines. The valley-head slopes above lowland stream-heads are occupied by relatively fine (gravel to fine sand) stratified deposits, which may be up to 10 m in thickness and grading downslope into alluvium.

2. Maliele loam or fine sandy loam. This is actually the only well-developed soil within this zone.

Pediment/Spur Slopes Topo-sequence.

The dissected, level to gently sloping plain below the escarpment and the scree slope, is characterised by three topo-sequences each related to the landforms and degree of dissection of the lowland area.

a. *Inter-spur Depressions Topo-sequence.* This consists of those soil series found on minor topographic depressions on the pediment slopes and spurs in the lowlands. It comprises mainly of two members and has a resemblance to Cave Sandstone Plateau topo-sequence.

1. Tsiki fine sandy loam on upper pediment slopes immediately below the scree slopes

2. Leribe fine sandy loam

b. *Spurs and Pediment Slopes Topo-sequence.* These are the soil series found on the broad spurs and upper pediment slopes

separated from each other by the drainage depressions of the dissected 1600 m surface. The typical members of this sequence are:

1. Leribe fine sandy loam, occupying the highest points on the spurs

2. Berea loamy fine sand on mid-slopes

3. Rama fine sandy loam

The whole of this topo-sequence is composed of soils derived from what has been referred to as Red Beds drift or aeolian sands. The sequence terminates above a minor escarpment at about 1670 m.

c. *Major River Depressions (Alluvial Flats) Topo-sequence.* The representative members of this sequence are mainly duplex soils and include the following soil series:

1. Khabo silt loam or loam to silty clay loam

2. Sofonia loam to light clay loam

3. Phechela silty clay loam, silty clay or clay

4. Maseru silt loam or loam to silty clay loam.

2.2.5 Vegetation and Land Use

The indigenous vegetation within Roma Valley and Maliele catchments is dominated by *Themeda-Festuca* grassland in the mountains and foothills regions, and *Themeda-Cymbopogon-Eragrostis* transitional to Highveld Sourveld grassland complex in the lowlands. Tree vegetation within the catchments is limited to planted groves of silver wattle, poplars, pines, eucalyptus and fruit trees around some settlements, and shrubs and some indigenous trees on the moist south-facing slopes. Some willows are found along the main streams in the area. The grasslands have been modified by cultivation and grazing. In some areas overgrazing has been so severe that the areas are being invaded by Karroid bushes and grasses (*Chrysocoma* sp.)

Land use in the area consists mainly of cropland, pasture and settlement areas,

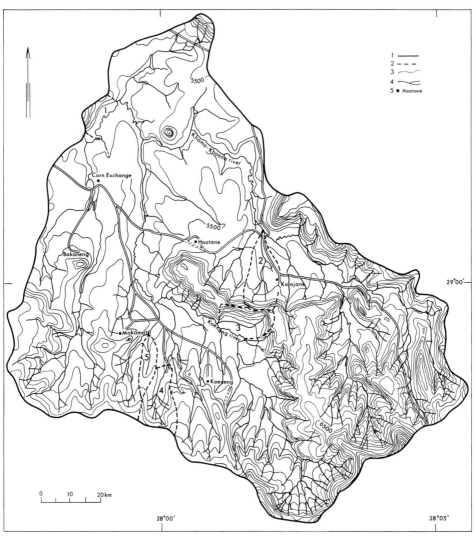

Fig. 2.15. Khomo-khoana catchment. Relief, hydrography and location of investigated subcatchments. **1**. Catchment boundary, **2**. Investigated subcatchments, **3**. Contour at 100 feet interval, **4**. Hydrography, **5**. Trading centres (stores).
Source: D.O.S. 421 (1:50,000, 1955—1966).

with the largest urbanization around the Roma Mission, St. Michael's Mission and the National University of Lesotho. Cropland is often open for grazing of the crop stub during the winter when all the crops have been harvested. Limited irrigation and fish culture are practised within the Roma Valley Agricultural Project Area.

2.3 Khomo-khoana catchment

Khomo-khoana catchment is a 5th order drainage basin in northern Lesotho and forms one of the minor subcatchments to the Caledon river basin (Figs. 2.15 & 16). The investigated catchment is exposed to the north-west and main streams flow in a north-westerly direction, generally falling

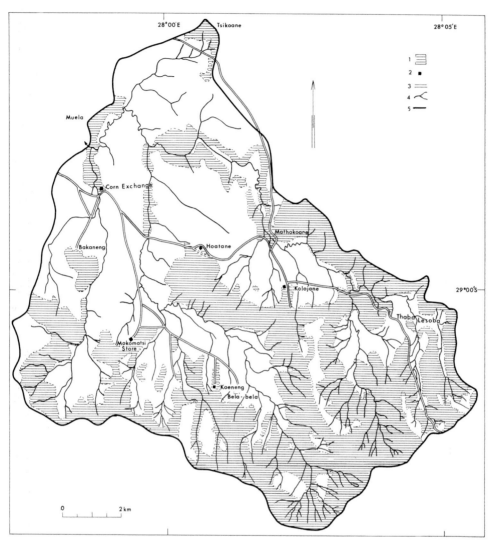

Fig. 2.16. Khomo-khoana catchment. Land use, hydrography and communications. **1**. Pasture, **2**. Trading Centres, **3**. Roads, **4**. Hydrography, **5**. Catchment boundary. The unshaded area represents cropland.
Source: D.O.S. 421 (1:50,000, 1955—1966).

from 2290 m to 1520 m. It is located at latitude 29° 06'—29° 53' S and longitude 27° 57'—28° 06' E and straddles the Leribe/Berea district boundary. The total area of the catchment is 163 km². The catchment has been a land management project area since 1973.

2.3.1 Bedrock Geology

The geological description of the catchment has never been made in any great detail. The information existing on the bedrock geology of the area is mainly based on the map by Stockley in his report on the geology of Lesotho (Stockley, 1947), and the works of Bawden and Carroll (1968), Binnie and Partners (1971) and the soil survey of the catchment by West

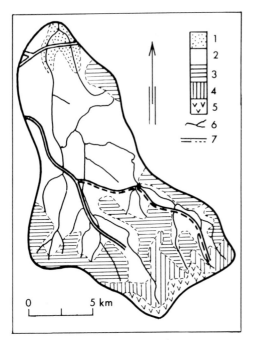

Fig. 2.17. Khomo-khoana catchment: Land systems. 1. Northern Beaufort plains, **2**. Red Beds Plains, **3**. Lowlands Escarpment, **4**. Northern Basaltic Foothills, **5**. North-west Escarpment, **4**. Hydrography, **7**. Roads.
Source: Carroll and Bawden, 1968.

(1972). To this should be added the present author's air-photo analysis and field observations on erosion and sedimentation in the catchment. The main body of the following description is based on these reports and oral information from geologists working in Lesotho.

From aerial photographs the lithological differences in the catchment can be easily observed and dykes are distinctly seen although some of them have been covered by recent deposits.

The geology of the Khomo-khoana catchment is similar to the geology of Lesotho in general and almost all the geological formations existing in Lesotho are represented in the catchment. Here the description follows the elevation of the catchment and the formations are described from the highest points in the catchment to the lowlands and ends with a brief description of the recent formations.

The catchment is underlain by sedimentary rocks and basaltic lavas of the Karroo System, Stormberg Series. The occurrence of the rock types corresponds approximately with the two physiographic provinces described under Section 2.3.3 below.

Drakensberg Beds

The highest parts of the catchment (1800—2290 m above m.s.l.) are underlain by the Drakensberg Beds, which are a series of basaltic lava flows, moderately strong, gray to dark gray and varying in thickness from a few centimetres to possibly 30 m. The lavas are stratified into parallel and almost horizontal layers interpreted as showing a mobile state at the time of deposition (Binnie and Partners, 1971). Pipe amygdales are normally found in a thin zone at the base of the individual flows and spherical amygdales at the surface of the flows. The dominant minerals are plagioglase feldspars, pyroxenes and some altered olivines (Cox and Hornung, 1966), other minerals reported are natrolite, heulandite, phyllipsite and apophyllite zeolite in the amygdales. Montmorillonite and vermiculite in finely disseminated form are reported present in the ground mass of the rock.

The beds are closely jointed at the surface, but this jointing does not reach any depth. Amygdaloidal basalts exhibit columnar jointing (Binnie and Partners, 1971, p. 14).

The boundary between lavas and the underlying Cave Sandstone formation is abrupt, tight and mostly unweathered where not exposed to the air.

Dolerite Dykes

Ridges and hollows of dolerite and gabbro contemporaneous with the lava flows are found criss-crossing the catchment. The direction of the dykes within the catchment is irregular, however, they are domi-

34

nated by the NE-SW and E-W major dykes with several minor ones having different directions. The dykes cut across both sedimentary and igneous rocks. Where they cut the sedimentary rocks, they are bordered by a baked zone of shattered and very closely jointed rock 0.5—3 m wide (Binnie and Partners, p. 16). The present author measured one such zone to be 5—6 m beside a dyke in the catchment along a foopath used for crossing the catchment boundary between areas 2 and 3. Most of the lowland dykes have been excavated for road materials and reservoir embankment filling.

Cave Sandstone

The Drakensberg Beds are underlain by Cave Sandstone and the contact zone follows approximately the 1800 m contour. The Cave Sandstone forms a single unit consisting of buff-coloured, weakly to moderately cemented, fine-grained sandstone forming cliff features with caves. The mineral composition of the rock is subangular quartz and feldspar grains with calcareous cement (Stockley, 1947; Binnie and Partners, 1971). The sandstone is easily weathered and where the surface is exposed there are small solution bowls up to 5 m in diameter and several pinnacles. It has poorly developed bedding planes commonly seen at the top and bottom of the formation. It contains lenses of clay-shales and is silty at the base. Binnie and Partners attribute the origin of the Cave Sandstone to aeolian processes because of the uniform grain size and lack of bedding planes (Binnie and Partners, p. 13). Jointing is rare but parallel joints may be observed along the cliff edge and these have been ascribed to stress relief and erosion of the underlying strata (Binnie and Partners, p. 13).

Transition Beds and Red Beds

Below the Cave Sandstone escarpment the catchment is underlain by sandstones, mudstones and clay-shales of Transition and Red Beds formations. These consist of red and buff sandstones alternating with purple, red and blue shales and mudstones. The mudstones and clay-shales weather easily and are responsible for the "terraced" form of the scree slopes. The differences between Red Beds and Transition Beds lie in the thickness of the alternating shales, mudstones and sandstones and in that the Transition Beds have grain size composition similar to that of the Cave Sandstone formation while the Red Beds are coarser in texture.

Recent Formations

These consist of alluvium, colluvium, wind-blown sands and products of weathering.

Alluvial deposits in the catchment are confined to narrow strips of silty materials bordering the major streams and are mostly found along the lowland streams. The erosion benches in the foothills are covered with a thin weathering layer which has been developed into soils.

On some of the foothill benches and remnants of the Cave Sandstone plateau there exists a thin cover of wind-blown soils consisting of silt and fine sand or direct products of weathering of the underlying sandstone rock.

Clay-shales weather rapidly under conditions of wetting and drying, producing cracks which allow access to water, the clay then crumbles and is washed away by sheet and gully action. This leads to undercutting of the mudstones and sandstones layers producing rock screes at the foot of the sandstone scarp. The basalt weathers to an ochre-coloured soil resistant to erosion by water, but the weathered matter is very thin on the mountain slopes.

Dolerite dykes weather in situ to thick, loose materials which may be up to 80 m thick (Binnie and Partners, 1971). This is used for road material and reservoir embankment filling. The exploitation of these weathering products has led to gullies and concentrated runoff from the hollows excavated.

In short, the bedrock geology of the catchment consists of basalt lavas, dolerite and gabbro dykes, sandstones, mudstones and clay-shales of the Karroo System, Stormberg Series ranging in age from Cretaceous to Triassic. These formations are overlain by varying thicknesses of recent formations in the form of alluvium, colluvium, weathering products and wind blown deposits (Table 2.1).

2.3.2 Climate and Hydrology

There are very limited data on climatic and hydrological observations within the catchment. Temperature observations are being taken at two stations which have been used to characterize the climate of the Khomo-khoana catchment. However, these stations are representative only for the lowland parts of the catchment. West (1972), in his soil survey of the Khomo-khoana catchment, described the climate of the area as follows: temperate, continental, subhumid with average annual precipitation ranging from 750 mm in the lowlands to over 900 mm in the mountains. More than 80% of this falls in the period October to April. Precipitation is erratic in amount, intensity and distribution both in time and space. The wettest and driest

months are January and June respectively. The winters have many subzero night temperatures, but rarely fall below –5°C, but may be lower in the mountains where snow is not rare. Mid-day in winter is normally sunny and warm. Normally, August and September are mostly windy with several light to moderate dust storms.

Temperature

Two observation stations can be used to estimate the temperature conditions in the catchment. These are Leribe Weather Station (297/083) and Maputsoe (296/745) located 10 km and 5 km north and west of the catchment respectively. The temperature data for Leribe station have been analysed by Binnie and Partners (1972) and are here reproduced as Table 2.5. The station has a temperature record of 15 years. They found the mean monthly temperature to vary from 7°C in June to 19.9°C in January. The annual mean is 11.8°C.

Table 2.5. *Monthly Air Temperature in °C at Leribe Station (after Binnie and Partners, 1972)*

Jan.	Feb.	Mar.	Apr.	May	June
19.9	19.7	17.9	13.9	10.7	7.0

July	Aug.	Sep.	Oct.	Nov.	Dec.
7.1	10.0	11.3	16.3	17.9	19.5

The Khomo-khoana Project has published the temperature data from Maputsoe station for the period 1974—1976 (FAO/TF/Les. 9) and these are represented in Fig.

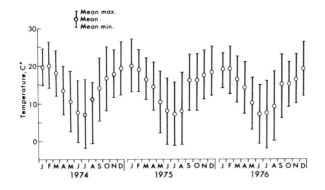

Fig. 2.18. Mean monthly and extreme temperature at Maputsoe, 1974—1976. *Source:* FAO/TF/Les. 9.

36

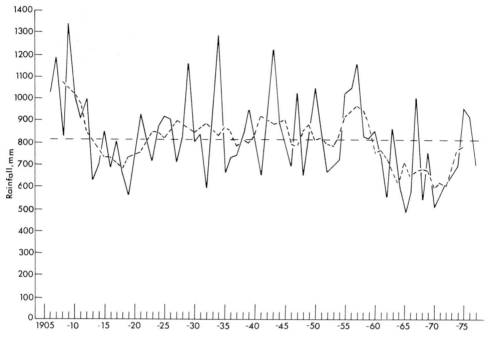

Fig. 2.19. Annual rainfall and 5-year running means, Leribe, 1906—1977. Solid line represents annual rainfall. Dashed horizontal line represents the mean for the whole period.
Source: Meteorological data to Sept. 1970. Climatological Bulletin, 1974—1977.

2.18 in the form of monthly mean maxima and mean minima. Diurnal temperature variations at Maputsoe are of the order of 10—12°C in winter and 14—15°C in the summer months.

Precipitation

Monthly rainfall totals have been published for Leribe covering the periods 1887 —1900 and 1906—1970 by the Ministry of Water, Energy and Mining, and for the period 1970—1976 in the Climatological Bulletin of the Hydrological and Meteorological Services of the same ministry (Fig. 2.19). Maputsoe data are published for the period 1972—1976 (FAO/TF/Les. 9, 1977).

In their description of precipitation of Lesotho, Binnie and Partners (1972) made the following remarks concerning the precipitation as recorded at Leribe: there are no data for rainfall duration less than one

day: 2-year maximum was calculated to be 50 mm per day corresponding to 6.1% of the mean annual precipitation; the calculated 50-year mean for the period 1920— 1970 was found to be 817 mm.

Streamflow

There are no data on streamflow in the catchment, but the general hydrology of the catchment can be described as follows:

1. *Perennial streams*—these are represented by the major streams up to second order creeks. The flows reach a maximum during the summer months of November to March, following the wet period. Minimum flows never reach absolute dryness for most of these streams except during the periods of extreme drought. They have flashy, short-lived peaks in their flows which often sweep off culverts and small bridges. These floods are erratic and depend on the rainstorm duration, rainfall

intensity, and the areal extent of the rainstorm.

2. *Intermittent streams*—most of the first order channels are fed by ground water overflows from springs during the wet period. The flow volumes of these streams are dependent on the precipitation amounts and normally run from after the first rains to the end of May and they are dry for the duration of the winter season. Most of these streams are deeply gullied drainage depressions with flows which rarely reach the main streams, but continue as subsurface flows through the deposits in the lower reaches of the depressions. To this group of streams should be included the small swamps and "lakes" found on top of the residual hills of Cave Sandstone and the depressions above the rest of the Cave Sandstone plateau.

3. *Ephemeral streams*—there are some streams, mostly first order and second order channels, where streamflow is limited to the short period during and immediately after rainstorms. They are represented by the lowland first order streams, overland flow along the topographic depressions, rain water springs and tunnel flows in the lowlands. In the basalt terrain, they are represented by most of the first order creeks and rain water springs emerging from the contact zones of alternating lava flows. During very wet years, they may carry water for the duration of the wet season, but seldom more than three months.

Summary and Discussion

All the data on temperature, rainfall and evaporation reported for the area should be modified to apply to the catchment. The rainfall is higher in the foothills and mountains and so is the humidity. This is shown by the short records in the Khomokhoana catchment run by the Project since 1976 and an old rainfall station at Koeneng store (Fig. 2.15). The temperature varies very much within the catchment depending on elevation, aspect and vegetation cover. Most of the south-facing slopes are both wetter and cooler than the open plains and the north-facing slopes. On these slopes frost lasts from April to August and mostly in the form of ice on the streams, pools and swamps below the slopes. Snow stays up to 3 months on the south-facing slopes in the mountain region, but as all the mountain slopes are north-west exposed in the catchment, snow cover has very short duration.

Rainfall within the catchment often occurs in bursts of high intensity and very short duration thunderstorms which seldom cover the whole catchment. The mountain region often experiences rainfall, leading to floods in the lowlands, while the lowlands do not. These floods are a great hazard to both human and livestock population in the lowlands. Several drownings have occurred in the lowlands as a result. These floods are very flashy in character and have very short durations, often lasting 10—15 minutes after the rainstorm.

The rainfall record for the periods 1906—1935, 1936—1965 and 1966—1977 is summarised in Fig. 2.20 together with the coefficient of variation.

The seasonal variations as shown by the monthly coefficient of variation have a maximum during the dry season. The 30-year annual means decrease from the 1906—1935 period as compared to the later periods by a total of 160 mm. This total decrease falls within the limits of error of the period of record. The analysis of the annual totals for the whole available record of precipitation shows that the high magnitude annual rains have decreased successively ever since the highest recorded value (1343 mm) in 1910 (Figs. 2.19 & 2.21). Three periods of drought can be observed from the 72-year record. The deepest of these deficits is the mid 1960s to 1974, reaching a minimum of the whole recorded rainfall in 1965 (482 mm). These yearly fluctuations have great impact on

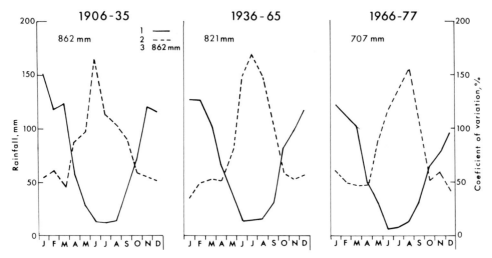

Fig. 2.20. Monthly and annual rainfall, and the coefficient of variation at Leribe for the periods indicated. **1.** Mean monthly rainfall, **2.** Monthly coefficient of variation, **3.** Mean annual rainfall for the indicated period. *Source:* Meteorological data to Sept. 1970. Climatological Bulletin, 1974—1977.

the vegetation cover and consequently the state and rates of erosion in the area. The largest erosion damage by water is usually during the wet years following immediately after two or more dry years. The largest changes in gully erosion in the catchment are associated with these periods (oral information from local villagers). This should be expected as most of the soils in the area are composed of expanding clays which crack during the dry period and these cracks easily form inflow channels into the underground channels and, when the rains come, they are already open to feed the subsurface channels and rapid slumping of the gully sides results, leading to both lateral and headward extension of the gullies or pipes.

Another aspect of rainfall worth noting is the occurrence of hail storms which are a great danger to crops and whose stones have a much higher detachment power on the soils than raindrops.

The lack of discharge and sediment load data in the catchment is a great drawback on the studies of the rates of erosion and sedimentation in the area. A rough estimate of these can be made by using the figures of runoff and sediment load

obtained by Binnie and Partners (1972) from the nearby Hlotse catchment upstream of Khanyane. These give a mean annual runoff of 180—200 mm and a mean annual sediment yield of 200—400 m³ km⁻². Future studies in the catchment are recommended to make use of a gauging station just downstream of the bridge over the Khomo-khoana river on the Maputsoe-Leribe road below St. Monica's Mission.

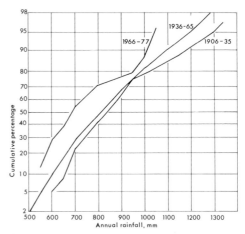

Fig. 2.21. Distribution of annual rainfall at Leribe for the periods: 1906—35, 1936—65 and 1966—77. *Source:* Meteorological data to Sept. 1970. Climatological Bulletin, 1974—1977.

This station had no record during the present studies.

2.3.3 Landforms and Surficial Deposits

The landforms within the Khomo-khoana catchment have been described generally in the works of Bawden and Carroll (1968), Binnie and Partners (1971), West (1972) and Chakela (1974). The last two reports deal with the Khomo-khoana catchment landforms specifically, while the other two deal with these as part of the general geomorphology of Lesotho.

In their work, Bawden and Carroll (1968), mapped four land systems in the area based on relief, landforms, soils and vegetation. These are summarised in Fig. 2.17.

In his report on the soil survey of the Khomo-khoana catchment, West (1972) gives the following summary of the geomorphological features and deposits found in the catchment: "river terraces, river floodplains, spurs and degraded spurs, sandstone scarps with scree and pediment slopes"... "The river terraces along the Khomo-khoana river were built up slowly and consist primarily of silt and very fine sand with some clay. There was practically no gravel in the upper 3 to 5 metres and pebbles were found in only few places in the upper part of the Foothills Region." The terraces have nearly level to gently undulating slopes, gentler away from the river, "resulting in backswamps... where the terrace meets the upland. Some of the backswamps have been partially or completely filled with recent erosion sediments" (West, 1972, p. 2).

Some of the significant geomorphological features not mentioned by Bawden and Carroll or West are the ridges and the depressions formed by dolerite dykes in the area.

Binnie and Partners (1971) give a more detailed description of landforms, although not specific to the area, but the main features present are related to the landscape genesis. Their description and my own field studies and air-photo interpretation are used in the description of the landforms and loose deposits in the area.

Binnie and Partners claim that the present landscape in Lesotho in general has been influenced by a series of changes of baselevel to which the rivers cut. All these changes have comprised a lowering of the base-level, interrupted by stable periods of varying duration. This has led to formation of several planation surfaces noted in the foothills and lowlands regions within the Khomo-khoana catchment.

A set of physiographic divisions has been used in all previous works on Lesotho, and to enable comparison with the other catchments in Lesotho, these are reproduced in Table 2.2 together with the land systems.

The following is an attempt at a more detailed geomorphological description of the catchment based on air-photo interpretation, topographic maps and field observation. The land systems of Bawden and Carroll (1968) are used as the main divisions of the catchment.

North-West Escarpment Land System

This land system forms the highest points within the Khomo-khoana catchment, and consists of mountains ranging in elevation from 2100 m to 2300 m, with steep simple slopes with north-east to south-west trend forming semi-circular valley-head slopes and dissected by several small streams forming a dendritic drainage pattern with several V-shaped valleys and spurs. The land system can further be divided into three land facets: steep, straight mountain slopes with terraces formed by layers of the lava flows; passes (wind gaps) of relatively flat or sharp-crested mountains at an elevation of 2100 m; and narrow valleys with rounded to sharp-crested interfluves containing first order streams, some of the streams carrying water only after rainstorms. Sheet erosion and rill erosion seem to be the dominant processes on these slopes and spurs.

Northern Basaltic Foothills Land System

The higher parts of this land system are on basalt, but the lower parts are underlain by Cave Sandstone formation (Land System 9, Bawden and Carroll, 1968) and its outliers in the form of isolated, flat-topped hills. The land system consists of an undulating basalt platform in which the streams have incised to varying depths often exposing large areas of Cave Sandstone, producing rounded spurs and spur-end slopes along the lower boundary of the land system. These spurs have stepped longitudinal profiles either determined by differences in weathering of the lava layers or by local changes in base-level. Within the Khomo-khoana catchment this land system lies between 1800 m and 1980 m contours.

The land system can further be divided into three landform units:

1. Flat to rounded spurs with convex side slopes and convex to straight and bluffy spur-end slopes.

2. V-shaped valleys separating the spurs reaching the sandstone in places but mostly on basalt.

3. Rounded and elongated depressions with springs and swamps, some of which have no permanent drainage channels. They seem to be cut into the Cave Sandstone formation.

The spur-end slopes and valley-head slopes of most of the minor streams in the land system are intensively rilled and most of the cultivated lands have lost a great deal of the surface layers, exposing raw mineral soils. The major gullies in this land system are confined to the depressions and the floors of the U-shaped minor catchments. The lower edges of the land system, along the Cave Sandstone edge, show stripping of the thin soils that have formed on them along a minor scarp 5—30 cm high.

Lowland Escarpment Land System

This land system is the most striking and almost dominates the landscape picture of the catchment. It is composed of two distinct landforms:

1. The Cave Sandstone escarpment which forms a line separating the lowlands from the uplands and consists of vertical cliffs, steep, stepped slopes or convex slopes.

2. Straight scree slopes with debris of pebbles and boulders separated by pockets of depressions with thick colluvial deposits at the heads of most of the lowland streams. Most of these depressional accumulations are heavily gullied to the bedrock. The lower ends of the scree slopes either terminate on pediment slopes or continue straight into the alluvial deposits of the main streams. The drainage channels either terminate in alluvial (colluvial) fans on the main river plane or as entrenched gullies forming tributaries to the main stream's entrenched reach.

Red Beds Plain Land System

The Red Beds Plain land system corresponds to the physiographic unit called the Northern Lowlands. It is composed of an undulating dissected plain stretching from the end of the scree slopes to the flood plain of the major streams. Three landforms dominate the land system:

1. *Pediment slopes*—short and narrow pediments at the foot of the scree slopes on the ridges separating the drainage depressions. These are mostly rocky and have very thin soil cover and sparse vegetation of short grasses and bushes. Most of the villages in the catchment are situated on this landform. Some of the pediments are found around dolerite ridges and at the foot of residual hills of the Lowland

Escarpment land system (Lowland escarpment outliers).

2. *Lowland spurs*—these are part of a surface gently sloping away from the pediments at a general elevation of 1650 m and are separated by entrenched river systems and gullies or depressions with sediment accumulation. They have steepened flanks and spur-end slopes. They are being lowered by wash and rill erosion as evidenced by a high concentration of rills, reduced vegetation cover and shift from their typical red soils cover to pale-coloured sandy soils.

3. *Alluvial flats*—a narrow and almost flat, branched zone along the main streams and their tributaries consisting of thick, dark brown, dark grey to black deposits. This forms the "minor river valleys" of Binnie and Partners (1972). The deposits in this zone have been incised by the streams, and along several reaches the streams never overflow their banks. In some reaches, the channels have such gentle gradients that deposition is taking place and streams have no defined channels.

The alluvial flats consist of two major types of landforms depending on the size of the stream or river along which they are developed:

a) The main alluvial plain with thick and relatively extensive deposits along the major streams. Here the streams have developed some terraces which are up to 3 m high, meander bars and cut-off meanders within the banks of the stream channels. Along some reaches, the streams are not entrenched and reed meadows dominate such locations, and alluvial fans have formed.

b) The tributary channel flats which are mostly occupied by large gullies with incision in the upper reaches and deposition in the lower sections, where the longitudinal profiles have lower gradients.

2.3.4 Soils

The soils of Khomo-khoana catchment have been described in details in the Soil Survey Report of the area by West (1972). Furthermore, some relevant information on the soils of the area is found in the broad groupings of the soils of Lesotho as mapped by Carroll and Bascomb (1967) and in volume 3 of Binnie and Partners' report on water resource development (Binnie and Partners, 1972b), covering the Lesotho lowland soils. The soil terminology used in the three works differs. In this report the terminology used by Binnie and Partners (1972b) is followed. Only soil groups are presented in this report, as they are more relevant to geomorphology and the processes of erosion and sedimentation under study.

Alluvial Soils

These are derived from river alluvium. They are dark-coloured, moderately stratified with varying texture and drainage capacities from well to poorly drained depending on their location. They are found on terraces of major streams, floodplains and pediment fans. Their texture is mainly fine sandy loams.

Duplex Soils

These are derived from younger colluvium or aeolian materials over older sediments. They are medium textured in the top horizons overlying fine textured colluvium or alluvium. The essential feature of these soils is "a light textured surface layer... overlaying a much heavier and usually older substratum" (Binnie and Partners, 1972b, p. 31). The transition between these two layers is very abrupt and an ashy-gray A2 horizon develops above it. Varieties of this exist where the textural succession is overturned. These soils are associated with gullying and piping. They occur in the floors of moderate to slight depressions and on pediment slopes.

Dolerite Soils

Two varieties of these soils have been described. They are both derived from basic, igneous rocks. They are well drained with texture varying from fine to medium and vary in colour from red to black and have a well-developed ABC profile. They occur on sandstone remnants in dolerite terrain, dolerite benches and ridges and dolerite footslopes and along drainage lines.

Red Soils

The origin of these soils is still disputed. Binnie and Partners (1972b) and P. H. Carroll (1976) assume that they are of aeolian origin while D. M. Carroll and Bascomb (1967) attribute them to Red Beds drift. The soils are red in colour, deep, and are found on broad convex spurs between 1600 m and 1750 m on what is thought to be an old erosion surface. They vary in texture from light to medium.

Sandstone Soils

These are mostly shallow—but sometimes deep—light yellowish-brown soils occurring on Cave Sandstone and Red Beds scarps and other sandstone outcrops that occur. They are characterised by illuviated, iron-rich, mixed materials, and formation of reddish-brown concretions. They are moderately to highly erodible.

Basalt Soils

These soils are restricted to the mountain province. They are reddish-brown in colour and have been referred to as eutrophic brown soils by Carroll and Bascomb (1967). Their texture varies from moderate to fine and they are very well drained with very low erodibility values.

The soils within the catchment show strong control by the geologic materials and topography of the catchment. Fig. 2.22 gives a schematic drawing showing this relation between soil series and geologic materials and topography as envisaged by West (1972). Table 2.6 names numbers of soil series given in Fig. 2.22.

2.3.5 Land Use and Vegetation

The Khomo-khoana catchment lies in the

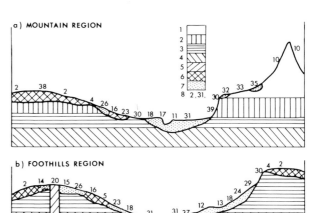

Fig. 2.22. Sketch diagram showing the relationship between: soil series, topography and geology within Khomo-khoana catchment as envisaged by West (1972). **1**. Drakensberg Beds, **2**. Cave Sandstone formation, **3**. Red Beds formation, **4**. Molteno Beds formation, **5**. Dolerite dykes, **6**. Quarternary and recent aeolian deposits locally reworked by colluvial action, **7**. Quarternary and recent alluvium/colluvium, **8**. Soil series numbers.
Source: West, 1972.

43

Table 2.6. *Soil Series Description. Key to Soils Series Numbers in Fig. 2.22 (After West, 1972).*

1. Leribe very fine sandy loam	20. Chaka gravelly very fine sandy loam and
2. Matsoete very fine sandy loam	lithosols on dolerite
4. Mathata very fine sandy loam	22. Kholokoe silty clay
5. Nyenye very fine sandy loam	23. Tuke very fine sandy loam
7. Matiki very fine sandy loam	24. Masaleng fine sandy loam
8. Maseru silt loam	26. Letsoela very fine sandy loam
9. Motlalehi very fine sandy loam	29. Nteleki association
10. Ralebese stony silt loam and associated	30. Lithosol on sedimentary rocks
11. Caledon fine sandy loam	31. Matsoali very fine sandy loam
12. Khabos silt loam	32. Lekhalong silt loam
13. Mohokare clay loam	33. Ntisa loam
14. Matšaba loam	35. Thabane silt loam
15. Chona Phantši loam	36. Bene clay loam
16. Bela-Bela very fine sandy loam	37. Kopi silt loam
17. Hlotsenyane loamy fine sand	38. Thoteng loamy fine sand
18. Likhetlane fine sandy loam	39. Mashoane very fine sandy loam

50. Severely gullied areas, stream banks and river wash

tall grass high veld ecological zone (West, 1972). According to Bawden and Carroll (1968), the catchment lies within the region of *Themeda-Cymbopogon-Eragrostis* grassland transitional to Highveld Sourveld. Table 2.7 gives a summary of the vegetation in the three physiographic units into which the catchment is divided (cf. Land Systems in Fig. 2.17).

Table 2.7. *Vegetation Types within Khomokhoana Catchment*

Physiographic Unit	Vegetation Types
Mountains	*Themeda-Festuca* grassland transitional to Highland Sourveld
Foothills	*Themeda-Festuca* grassland transitional to Highland Sourveld.
Lowlands	*Themeda-Cymbopogon-Eragrostis* grassland transitional to Highland Sourveld

The vegetation is mainly tall grasses and herbaceous weeds and shrubs mainly on the south-facing slopes. Tree vegetation in the area is limited to small groves of trees around the villages, and some groves of indigeneous trees and shrubs on the wetter south-facing slopes. The absence of tree vegetation throughout the country has been explained as due to frost, droughts, fires and overgrazing.

The land use is rather complicated. During the summer and autumn the land can be divided into cropland, pastures and village areas, but during the winter most of the cultivated lands are opened to grazing of the crop stub by livestock. Large parts of the area around the scree slopes and pediment slopes are lands normally meant for grazing and settlements. Steepness of the slope does not seem to be any criterion for the cultivation and this has led to severe surface wash and gullying on the upper, steeper slopes in the catchment especially in the foothills and mountain provinces.

3. Methods

3.1 Introduction

The aims and purposes of these studies are outlined in Chapter 1 of this report. In this chapter, the methods used in these studies are presented. The methods can be conveniently put into three groups, namely, reservoir surveys, catchment surveys and measurement of water and sediment discharge. All the three groups of methods were applied in the studies of catchments situated within the Roma Valley and Maliele catchments. In the Khomo-khoana catchments only catchment surveys were employed.

3.2 Reservoir surveys

Reservoir surveys were undertaken in 6 reservoirs located within the Roma Valley and Maliele catchments (Table 3.1).

The procedure for the reservoir surveys was as follows. A grid system was established for each reservoir in the following manner: a base line was established along the reservoir embankment as close to the full supply level as possible. The extreme ends of the base line were permanently marked with iron rods or permanent marks of paint on the concrete spillway walls. A distant point was selected whose line of sight from the extreme ends of the base line formed an angle as close to a right angle as possible. A line was then marked on the ground in alignment with the distant point and marked at 10-metre intervals towards the inflow channel of the reservoir. Where no distant point could be located, a compass sighting was made at a right angle to the base line or reservoir embankment.

The establishment of the line at right angles to the base line was made from both ends of the base line, thus making a rectangular coordinate system over the reservoir. The base line was then divided into 5-metre intervals and marked with wooden pegs. In this way, a 5×10 metre grid system was established for each reservoir.

The left end-point of the base line was designated as the zero point. The reservoirs were then ready for bottom sounding.

The points for bottom sounding were located by strectching a thin, graduated wire, marked at 5-metre intervals, between the endpoints of lines parallel to the base line (reservoir embankment) and 10 metres apart. The wire was tied in such a way that an even metre point coincided with the reference line at right angles to the base line. The depth was measured at the 5-metre points using a hollow aluminium rod, 2.5 cm in diameter,

Table 3.1. *Roma Valley and Maliele Catchments: Reservoir data 1973/74.*

Reservoir No.	Location		Catchment area, km^2	Reservoir area, m^2	Maximum depth, m
1	Roma Valley	area 1	26.0	4163	3.0
2	”—	2	0.5	12700	5.0
3	”—	3	0.6	20307	4.5
4	”—	4	2.2	27490	6.2
5	Maliele	area 1	0.6	9600	5.0
6	”—	2	6.3	6515	1.4

graduated in centimetres. The readings were made from the water surface and read to 1 centimetre. The readings were later reduced to depths below full supply level of reservoir. The error in depth readings is of the order of 1 cm due to the influence of the boat and the waves. On windy days this error could be up to 3 cm. During the sounding it was necessary to ensure that the same points were measured during the future surveys, and that the line of measuring was the same. To accomplish this, two field assistants stood on either side of the line and kept the boat in line at the time of sounding. This reduced the side error in the location of the sounding points to within 1 metre, depending on the bottom topography of the reservoir.

At the completion of the bottom soundings in 1973/74, the shoreline was surveyed with level tube/theodolite and staff level. The end-points of the sounding lines were surveyed into the shoreline map and pegged to ensure location of sounding lines for future measurements. The reference distant points and compass lines were surveyed with a theodolite in order to determine exactly the angle between the base line (reservoir embankment) and the longitudinal lines of the reservoir grid system.

The bottom soundings were repeated in the 1975/76 rainy season for some reservoirs or parts of the reservoir, and in 1977 for all the reservoirs.

The data obtained from the reservoir soundings were used to produce bottom maps of the reservoirs with 0.5 m or 1.0 m contour intervals. The volumes of the reservoirs were computed from the bottom maps as follows: the area enclosed by each contour line was determined by planimeter and the volume of the reservoir computed using the formula

$$V = 0.5\, h \sum_{1}^{n-1} (A_i + A_{i+1}) + d_n A_n \qquad (3.1)$$

where V is the total volume of the reservoir in m^3, h the contour interval in m, A_1 the

surface area of the reservoir at full supply level, A_n the area enclosed by the lowest contour, and d_n the mean depth of measured points below the lowest contour.

The longitudinal line of maximum depth was chosen for each reservoir and bottom core samples were taken with a gravity corer (Fig. 3.1). The samples were visually inspected for structure and hand-felt for texture characteristics. Some of the samples were kept for laboratory analysis of grain-size. The core samples taken during bottom sounding in 1977 were weighed in the field with a spring balance and their wet bulk densities determined. The samples were then transferred to metal containers and transported to the Department of Geography, the National University of Lesotho, where they were air-dried and crushed. Grain-size analysis was made by sieving and the fractions less than 0.063 mm were collected and kept in numbered plastic bags and later transported to Uppsala where pipette analysis was performed. Particle density and loss of weight on ignition were also determined on some of the samples in Uppsala.

In connection with reservoir surveys, sediment samples were collected from the sandbank deposits at the inflow and on the side of the reservoirs for the study of composition and types of the deposits. The samples were treated like the core samples described above. The majority of sandbank samples were, however, only fractionated by sieving.

In summary, the reservoir surveys included bottom sounding of the reservoirs, sampling of reservoir bottom deposits for grain-size analysis and study of sedimentary structures and determination of bulk density of the sediments.

The difference in reservoir volume between the successive years represents the amount of sediment deposited in the reservoir during the intervening period.

The sediment accumulated in the reservoirs is only a portion of the sediment delivered to the reservoir by the feeding

Fig. 3.1. Gravity corer (from Axelsson, 1979).

1979, Ashida, 1980) and then used to relate reservoir deposition to sediment yield of the catchment.

In the present studies, deposition of suspended sediment in the reservoirs is estimated by the sedimentation formula worked out by Sundborg (1956, 1958, 1964b) based on the consideration of reservoir deposition in relation to the physical processes of sedimentation. The formula has been applied with good results in predicting river, reservoir and lake sedimentation (Sundborg, 1963, 1964a; Axelsson, 1967; Nilsson, 1974 & 1976). The formula was used to estimate the amount of sediment likely to be deposited in the reservoir under given flow conditions and material properties of the streams feeding the reservoir. Karaushev (1966) developed a similar formula for siltation of small dams. The modified form of the Sundborg formula used in these studies can be expressed as follows for any specific grain size.

$$S_{mo} = S_{mi} e^{- \frac{c \cdot f(c) \cdot L \cdot B}{Q}} \qquad (3.2)$$

where S_{mo} is the average concentration of suspended sediment of a specific grain size at the outflow section of the reservoir, S_{mi} the average concentration of the same grain size at the inflow section of the reservoir, L the length of the reservoir, B the width of the reservoir participating in through flow, Q the water discharge at the inflow section of the reservoir, c the settling velocity of the specific grain size, and f(c) a function which gives the relationship between the mean concentration of the specific grain size near the bottom and the mean concentration of the same grain size in the vertical.

Another estimate of trap efficiency was made using the method developed by Churchill (Strand, 1975) whereby the sedimentation indices of the reservoir are established and used for estimating the trap of efficiency of the reservoir.

streams, the rest is transported across the reservoir to the lower reaches. The ratio of the quantity of sediment deposited in the reservoir to the total sediment inflow into the reservoir is known as the trap efficiency of the reservoir (Chow, 1964, Strand, 1975, Gill, 1979). Trap efficiency of a reservoir can be estimated by use of theoretical and empirical relationships between the characteristics of the reservoir and the physical properties of the catchment area upstream of the reservoir (Sundborg, 1964a, Strand, 1975, Gill,

Churchill's sedimentation index (Strand, 1975) is a function of the capacity of the reservoir, length of the reservoir, the mean inflow rate into the reservoir, the period of retention, and the mean water velocity across the length of the reservoir. Sedimentation indices derived from this relationship are then used to read off the percentage of silt likely to be retained in the reservoir at a given index from a graphic relationship between indices and percent incoming silt passing through the reservoir. Trap efficiency estimates were made only for those reservoirs which have perennial feeding streams and have long periods of spill-over (reservoirs 1, 4 and 6).

3.3 Catchment surveys and gully extension measurements

The purpose of the catchment erosion surveys was to locate areas of erosion and deposition within each study area and to distinguish erosion types and, where possible, their intensities and spatial distribution.

Three methods were employed in these studies for the survey of erosion and sedimentation.

1. Air-photo interpretation after reconnaissance field checking in 1973.

2. Field inventory of erosion and sedimentation mainly using slope traverses from valley bottoms to the water-divide.

3. Repeated surveying of cross-sections of gullies, annual checking of bank erosion and gully slumping, and measurement of gully-head extension using bench marks (pegs) installed at some distance above the gully-head on some side gullies, and those main gullies that had not reached the bedrock scarp or another form of obstacle for headward extension.

The materials used for air-photo interpretation were aerial photographs with a general scale of 1:30 000 and 1:28 000 taken in 1951, 1961 and 1971; the photo mosaics and ortho-photo maps at the scale of 1:10 000 made from the 1971 photographs; contoured maps at the scale of 1:50 000 compiled from the 1951 aerial photographs. The aerial photographs were inspected stereoscopically using mirror stereoscopes with 3x and 8x magnification, locating types of erosion and cultivated areas. The information was then transferred to traced sheets of photo mosaics/ortho-photo maps and the gullies numbered for each of the detailed study areas. A scale was worked out for each numbered gully using the air-photographs and the maps. A group scale was calculated for a group of gullies located close to one another and equal distances from the centre point of the air-photograph. This was done for only those gullies that showed differences in gully lengths, form or size between 1951 and later. The gully lengths were measured on the photographs using a magnifying tube with a millimetre scale and 7x magnification. The readings on the tube scale can be read to 0.05 mm. The readings were then transformed into actual distances using the calculated scales. The same was done subsequently for the photographs of 1961 and 1971. The differences between the lengths of the same gully between the different dates give the gully extension. Because of the uncertainties in the scales it was thought wise to measure only those gullies that showed more than 4 metres in headward extension. The gullies showing apparent changes in form but were less than 4 metres in headward extension were recorded as showing probable lengthening.

Rill, sheetwash, homesteads and areas of exposed bedrock were delineated and marked on mosaics/ortho-photo map sheets to give an areal pattern of the distribution of these processes. The aerial photo information was continually checked and adjusted during field studies and complemented with the information about other processes in the areas such as slides, pipes and fossil slides, which showed only as

48

depressions in the terrain on the aerial photographs.

The sites for slope traverses were preliminarily selected by interpretation of 1961 air-photographs and analysis of 1:50 000 maps. For the slope traverse studies, lines were chosen from the stream channel to the top of the scarp or to water divide or along a side gully of the main drainage channel from the confluence with the main channel to the water divide and a little beyond. Sometimes the profile was terminated at the bottom of a cliff scarp.

The slope profile surveys were made using a hand clinometer (Meridian), a staff (marked at the observer's eye-height), and a 50 m steel tape. The slope measurements were made at 50-metre intervals (less where there was a change in slope gradient within the 50-metre section). The observations made for each 50-metre section were: slope gradient in degrees and read to the nearest 0.5°, vegetation type and cover, land use type, dominant erosion features, and soil type. Where the slope profile was measured along a gully, the depth of the gully was measured and cross profiles were made at arbitrary intervals. The grass cover measurements were made along a line by measuring the length of bare ground along the line and converting this into linear percentage grass cover for the section.

Cross profile lines were established in some gullies to measure the sedimentation and erosion along the gully floors and for measuring lateral extension of the gullies. These were re-surveyed in 1975 and 1977. Longitudinal profiles of the main gullies in all the study areas were made and the sections showing deposition and erosion noted.

In connection with the longitudinal profile surveys of the main gullies, samples of the deposits in the gullies were taken and grain-size analyses of these were made. The morphology of the deposits was studied in the field and the description kept together with the sample numbers of the deposits.

Pegs were installed at gully-heads of some side gullies and those main gullies that had not grown fully. The distances between the pegs and the gully-head were measured occasionally and annually. The occasional measurements were made after heavy rainstorms. On each measuring occasion the following parameters were measured.

1. Gully-head to peg distance
2. Depth at the gully-head
3. Width of the gully 1 metre downstream of the gully-head.

Observations were made of any sign of slumping, cracking and moisture conditions at the base of the gully-head scarp.

During each rainy season the main gullies were inspected at least once from the lower end of the gully to the gully-head for lateral erosion and slumping. The slumped blocks or scars they left were measured by recording their lengths, maximum width and the depth of the scar/block. These data were then transferred to the overlays of either air-photographs or traced sheets of the mosaics/ortho-photo maps. Several profiles were described in connection with these annual surveys.

A complementary inventory of mass movements in the catchments was carried out in the 1977 rainy season. This included location and description of rockfalls, slumps, and slides on the hillslopes. Two major slides which occurred after the November rain in the Roma valley and one in Maliele catchment were surveyed. The location of fossil slides and slumps were recorded for the upper Roma valley and Khomo-khoana area 3.

3.4   Water and Sediment Discharge

Staff level gauges were installed in all the 6 reservoirs in 1973/74 and daily readings made to supply data on the storage fluctuations of the reservoirs. The record for

49

these observations is not complete and it is only limited to those months the writer was in Lesotho as it became difficult to find personnel reliable and willing to work continuously over the whole research period. Therefore the storage data cover parts or whole of the rainy seasons 1973/74, 1974/75, 1975/76 and 1976/77, the longest sequence of data being for the second and the fourth rainy seasons. Occasional water sampling was undertaken for sediment concentration analysis of water flowing into and out of the reservoirs at high flows.

Three river stations were selected in 1973 along the Roma Valley and in Maliele catchment. The Roma Valley river stations are located at St. Joseph's bridge (near the St. Joseph's Hospital in Roma above reservoir 1), and at Lehoatateng bridge, and forms the outlet section of Roma Valley catchment 5. The Maliele station is located at the new bridge at the outlet of Maliele catchment at Mahlabatheng. The cross-sections of these stations were surveyed and staff level gauges installed for manual reading in 1974. Like the reservoir storage gauges, the observations for these stations are limited only to those periods of my visits in Lesotho.

For the first two rainy seasons (1974/75 and 1975/76), water levels were read daily in the morning hours for all the river stations. At the end of the rainy season in 1976, all the staff gauges had been destroyed through vandalism, and as they were not tied to any permanent bench marks, their record is not comparable with that of the 1976/77 rainy season. In October 1976 new water level gauges were installed at the three river stations and two field assistants put in charge of the daily water level readings for the period 28 October 1976 to 15 April 1977. The observations were made in the morning hours between 10 and noon. The three river stations were rated five times during the period by wading using a propeller current

meter (Ott Pigmy). At each rating the sections were surveyed to check changes in the cross-sectional form. The current meter measurements were made at intervals of 1 metre across the stream and where the water depth was less than 50 cm only one reading was made per vertical at 60% of the water depth, otherwise three readings were made per vertical at 80, 60 and 20% of the depth below water surface. Discharges were calculated for the measurements and discharge rating curves established for each station. These data are dealt with under relevant sections in Chapter 4.

Occasional water sampling for the analysis of sediment concentration was started in the 1974/75 and 1975/76 rainy seasons at the three stations. The sampling was carried out using a time integrating hand-operated sampler (Fig. 3.2, cf. also Nilsson 1969) and 1-litre bottle. The water sampling programme was intensified after the installation of new staff levels in the rainy season 1976/77. The intensified programme involved collection of water samples at each river station three times a week and on rainy days which did not coincide with water sampling days. The samples were collected at points of maximum turbulence in the section. At high water the sampling depth used was 60% of the depth below the water surface, at low water the samples were taken in such a way that the intake nozzle of the sampler was at least 6 cm below the water surface.

The determination of sediment concentration in water samples collected during the 1974/75 and 1975/76 rainy seasons was carried out at the laboratory of the Chemistry Department, National University of Lesotho (N.U.L.) as follows: the water samples were transferred from the sampling bottles to graduated glass beakers and the volume read off. 2—3 drops of 1N HCl were added and the beakers allowed to stand until all the sediment had precipitated leaving a clear supernatant liquid above. As much as possible of the

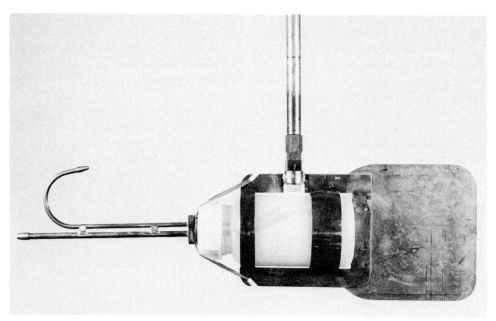

Fig. 3.2. Water sampler (from Nilsson, 1967).

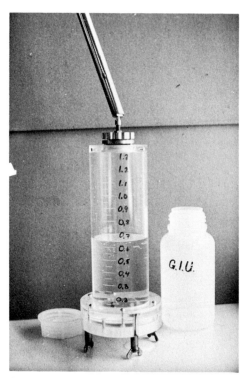

Fig. 3.3 Water pressure-filtering cylinder. (Photo: J. Rehn, 1980.)

supernatant liquid was decanted and the rest transferred to clean, weighed beakers and dried overnight in an oven at 105°C. After cooling, the beaker and sample were weighed and the weight of sediment determined by difference. This weight was then divided by the volume of the water sample to give sediment concentration in $g \, l^{-1}$.

The water samples collected during the intensive water sampling programme of the 1976/77 rainy season were either filtered directly in the field, or later at the Department of Geography, N.U.L., using a pressure-filtering set (Fig. 3.3). The filter papers used were Munktell filter paper No. OOH and Whatman No. 542. The filter papers with the sediment were folded and kept in numbered plastic bags for transportation to Uppsala. The filtrate was measured and the volumes recorded and kept together with the filter paper numbers and dates of sampling. All the filter papers were ignited at 550°C in the laboratory of the Department of Physical

Geography, Uppsala, and dry weights of the samples determined. Sediment concentration was then calculated by using the recorded volumes and the dry weights of the samples after ignition, and expressed in mg l^{-1} inorganic suspended sediment.

Water stage, water discharge and sediment concentrations and load data for each river station are presented under relevant sections of this report in Chapter 4.

4. Soil erosion and reservoir sedimentation: Roma Valley and Maliele catchments

4.1 Roma Valley area 1

4.1.1 Introduction

The catchment forms the upper reaches of the Roma Valley catchment described in Chapter 2. The investigated subcatchment lies upstream from the N.U.L. intake reservoir (Figs. 2.1, 4.1, 4.3 and Table 6.3). The relief of the catchment varies from 1670 m to 2504 m above m.s.l.

In order to simplify the description of erosion features and present day processes active within the catchment, the land systems presented in Chapter 2 have been subdivided into smaller landform units based on landforms, relief, geology, hydrology and vegetation. In order of decreasing altitude, these landform units are described below.

Mountain Slopes

The headwaters of the drainage area consist of steep, straight to slightly concave slopes showing terraces of different lava flows. These mountain slopes are drained by several intermittent and ephemeral streams and a few perennial streams, forming minor valleys separated by rounded to crested interfluves. The tops of the highest mountains in this area form an almost level surface at approximately 2500 m. The surface slopes slightly to the southwest.

Basalt Foothills and Cave Sandstone Plateau

The dominant landform in this zone is a structural platform of Cave Sandstone forming an undulating plain with broad, rounded spurs separated by minor stream valleys. The other two landforms of significance in the area are the basalt planation surface at approximately 2200 m and ridges of dolerite dykes. The two planation surfaces in this landform grouping are separated by a minor scarp of basaltic lava. The main river and its tributaries have cut into the underlying bedrock to varying depths producing deep to shallow gorges (Figs. 4.2 & 2.10). A fourth, not easily noticed geomorphological feature in this zone, particularly on areas underlain by sandstone, is a spoon-shaped depression filled with loose deposits and formed on the Cave Sandstone formation.

Cave Sandstone Escarpment and Colluvial/Scree Slopes

At an elevation of about 1680—1700 m, the slopes steepen upwards and have gradients of 20°—35°. The zone on either side of the centre of the catchment is covered with colluvial material strewn with large boulders on a series of cones forming lobes down the slopes, with boulder size increasing downwards (Figs. 2.13, 2.14). The colluvial slopes terminate upwards under a vertical cliff of Red Beds formation capped with Cave Sandstone, the top of which has a general elevation of 1800 m but may be higher. The cliff face is marked by either broad bluffs with or without caves, hollows where the dolerite dykes cross the Cave Sandstone formation, or minor gorges with falls or rapids (Fig. 2.11).

Pediment Slopes and Spurs

On either side of the river valley flats is a

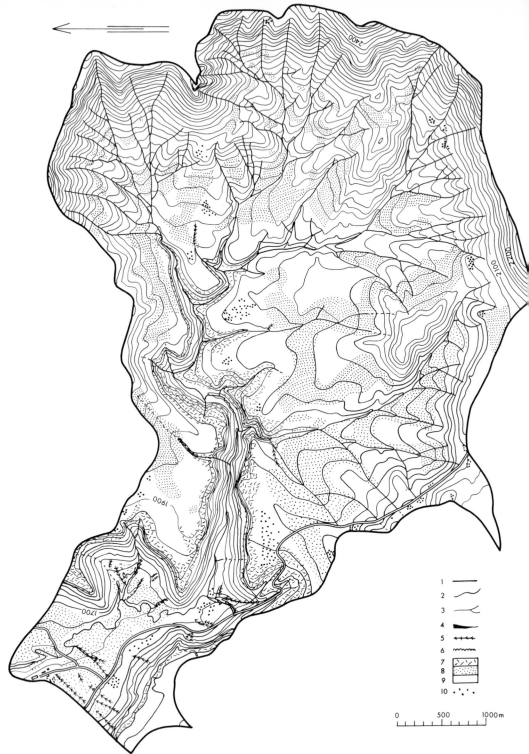

Fig. 4.1. Roma Valley Area 1: Land use, hydrography relief and gully erosion. **1**. Catchment boundary, **2**. Countours at 100 m interval, **3**. Hydrography, **4**. Investigated reservoir, **5**. Gully, **6**. Rill, **7** Rocklands, **8**. Cropland, **9**. Pasture, **10**. Homesteads.
Source: Thaba Bosiu Rural Development Project's ortho-photo maps (1:10,000, contoured) and airphotographs, taken in 1951/52, 1961/62 and 1971.

Fig. 4.2. Maphotong gorge (Photo: Q. K. Chakela, April, 1977).

sandstone terrace forming a gently sloping surface covered with relatively thin deposits (origin uncertain but probably a mixture of alluvial and aeolian deposits) and extends to the foot of the steep slope upwards. This surface is more extensive on the western side of the river and has a concave downward profile with steepened lower slope. On the eastern side of the main channel it consists of a narrow sandstone bench. It forms most of the upstream part of a dissected, gently sloping surface claimed to be an old erosion surface (Binnie and Partners, 1971). The river valley flats, pediment slopes and spurs form the Central Lowlands Land System of Bawden and Carroll.

River Valley Flats

The lowest parts of the catchment from the area near lower Maphotong village to the reservoir at the outlet of the catchment, consist of a narrow zone of alluvial deposits, with a meandering channel cutting through these dark gray to black and thick

deposits to maximum depths of 15 m in places. The slope of the river plain is almost level to gently sloping and steepens upwards and downwards from the centre of the reach through a series of secondary scarps and minor rapids in the main river. Small river terraces have been built at some sites in this reach but they are very insignificant in comparison with the general morphology of the valley flats. The zone is approximately 200 m at the widest and only 30 m at the outlet. It passes into the uplands through a gorge within which it diminishes to the size of the entrenched channel 2—10 m wide.

Land use and Land Management Practices

The land use in this catchment can be divided into three major categories: cultivation, grazing and settlements (villages). The largest part of the lowlands in this catchment is fully used for cultivation of the main staple crops, i.e., maize, sorghum, beans, peas and to a limited extent winter wheat. These fields are opened to

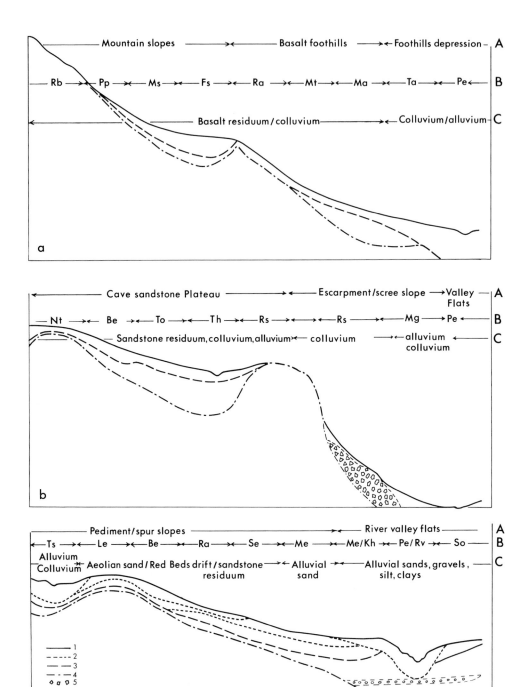

Fig. 4.3. Sketch diagram of topo-sequences of Soil Series within Roma Valley and Maliele catchment. A. Landform zones, B. Soil Series symbols (see table 4.12 for explanation), C. Surficial deposits. **1**. Surface profile, **2**. Pronounced horizon boundaries within the soil profile, **3**. Soil-surficial deposits interface, **4**. Surficial deposits—Bedrock interface, **5**. Scree boulders, **6**. Gravels and very coarse sands.
Source: Carroll, P. H. et al. 1976.

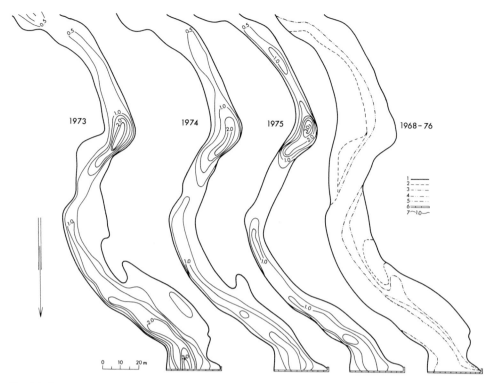

Fig. 4.4. Bottom maps of reservoir 1, and shoreline changes between 1968 and 1976. **1.** 1968 shoreline and full supply level, **2.** 1973 shoreline, **3.** 1974 shoreline, **4.** 1975 shoreline, **5.** 1976 shoreline, **6.** Reservoir embankment/crest overflow spillway, **7.** Contour in m below full supply level.

grazing of the crop stubble after harvest and remain so for the greater part of the dry season (June—September). The upper foothills and mountain slopes area are mainly grazing lands, as also are the colluvial slopes. But limited cultivation is practiced here leading to cultivation on very steep slopes with the result that erosion processes have become activated in the zone. The Cave Sandstone plateau area is mainly for cultivation with grazing limited to areas between the fields, along the lower edges of the plateau, spur slopes, hillcrests and around homesteads.

Most of the villages are situated around hillcrests in the foothills region, just below and above the escarpment/scree slope areas in the lowlands. The biggest villages are those around the Roma Mission area and along the road through the catchment to the highlands. The area below the Cave

Sandstone escarpment has been a project area for land development and improvement of agriculture by the Roma Valley Agricultural Project since 1965. The activities of the project include: limited irrigation farming; cooperative mechanized farming; combating of gully erosion through construction of dams, rockfills in gullies and construction of stone walls on some actively eroding gully-heads; improvement of communication routes in the area to facilitate the movement of farm machinery. The whole catchment is part of Thaba-Bosiu Rural Development Project. Improvement of water supplies in the catchment consists of the N.U.L. intake reservoir which supplies water to the university area situated just outside this study area. The reservoir is studied and described in Section 4.1.2 of this report.

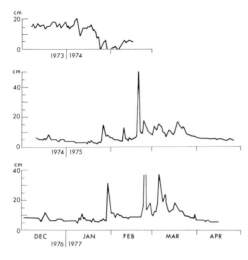

Fig. 4.5. Water level at outflow section of reservoir 1 during the rainy seasons: 1973/74, 1974/75 and 1976/77.

4.1.2 Reservoir Surveys

Reservoir 1 forms the lower boundary of Roma Valley area 1. The reservoir was completed in 1968. The original depth of the reservoir at the dam wall was 12 m. The depth was reduced to less than 3 m between 1968 and 1970 when a further 3 m stone wall was added. In 1973 the reservoir had been reduced in depth to 3 m, a loss of 3 m in 3 years. The width of the reservoir was also severely reduced between 1968 and 1973 (Fig. 4.4).

Reservoir surveys were made in the rainy seasons 1973, 1974/75 and 1975/76. The surveys included: bottom sounding of the reservoir and production of bottom maps, occasional sampling of water at the inflow and outflow sections of the reservoir for analysis of sediment concentration (Fig. 4.15 and Table 4.1), sampling of sediment from the sandbanks and bottom of the reservoir for the analysis of texture of reservoir deposits (Fig. 4.6b), recording of water level in the reservoir to provide an index of reservoir storage variations (Fig. 4.5).

The bottom maps of the reservoir were used in computing the reservoir volumes for the three periods (Table 4.2). The differences in reservoir volumes represent the amount of sediment deposited in the reservoir in the period between the surveys. The volumes of the reservoir deposits were found to be 1240 m³ between 1973 and 1974, and 400 m³ between 1974 and 1975 giving a total sediment accumulation of 1640 m³ between 1973 and 1975. The reduction in sediment accumulation in the second period is a sign of the rapid decrease in trap efficiency of the reservoir as it developed from a pool to an ordinary low gradient channel reach between 1973 and 1976.

The median and mean (Folk and Ward, 1957) grain sizes of the reservoir deposits

Table 4.1. *Sediment Concentration at the Inflow and Outflow Sections of Reservoir 1, 1973—1975.*

Date	Stage	Sediment concentration in mg/l	
	cm	at inflow	at outflow
73-12-16	12.8	2560	2110
24	14.3	200	100
30	17.8	10	10
74-01-03	15.0	320	340
08	13.5	470	940
09	7.5	370	4120
23	0.0	2320	2280
25	3.0	620	500
75-01-24	not measured	71	84
27	”—	500	310
02-10	”—	1167	378
21	”—	—	22
24	”—	1056	—
03—05	”—	—	62

Table 4.2. *Reservoir Volumes and Sediment Accumulation, Reservoir 1.*

Date	Reservoir area at F.S.L. m²	Reservoir volume at F.S.L. m³	Sediment volume m³
December, 1973	4,163	3,472	
			1,238
December, 1974	3,178	2,234	
			404
December, 1975	2,518	1,830	
Total sediment accumulation: 1973—1975			1,642

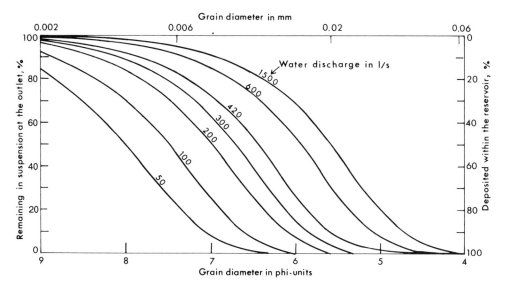

Fig. 4.6a. Relative estimates of trap efficiency of reservoir 1 based of Sundborg formula. The calculations are based on the following assumptions: a) 1975 area is taken to be participating in throughflow, b) A water temperature of 20°C, c) No resuspension deposited sediment, d) Sediment particle density of 2.65 gcm^{-3}.

are 0.11 mm and 0.15 mm respectively (Fig. 4.6b) with the material finer than 0.063 mm making up 10—40% of the reservoir sediments. This implies that a large part of the fine sand, silt and clay size particles are transported across the reservoir by most flows except the low flows during the dry season.

The sediment trapped by the reservoir corresponds to 30 t km^{-2} y^{-1} of the sediment yield from the catchment. This value is low in relation to the probable denudation rate of the catchment. During the rainy season 1976/77, at the river station load of 230 t km^{-2} was observed (cf. Section 4.1.4). This supports the conclusion that a greater part of the sediment reaching the reservoir as suspension load is transported through the reservoir to the lower reaches.

The relationship between the sediment reaching the reservoir and that trapped in the reservoir is given in Fig. 4.6a and Table 4.4. The relationship illustrated in Fig. 4.6a is based on calculations using the Sundborg formula discussed in Chapter 3 for some discharges measured at the river

station upstream of the reservoir during the rainy season 1976/77 and assuming that the water temperature was 20°C and that the 1975 surface area of the reservoir was involved in through-flow. Even at the low discharges used in this calculation, all material finer than 0.063 mm is shown to be carried across the reservoir in suspension, implying that since 1973 the greater part of sediment reaching the reservoir leaves the reservoir during moderate to high discharges and deposition occurs only at very low flows. The trap efficiency estimate made using the Churchill method (Table 4.4) for inflow rates represented by the mean and median discharges measured at the river station upstream of the reservoir and reservoir volumes of 1973, 1974 and 1975 varies from zero to 50% for silt-clay materials. The estimated sediment yield for the catchment becomes 20 m^3 km^{-2} y^{-1} and 300 m^3 km^{-2} y^{-1} for the median and mean discharge conditions respectively.

The sediment yield estimates obtained above should be considered in relation to the reservoir storage, and the amount of

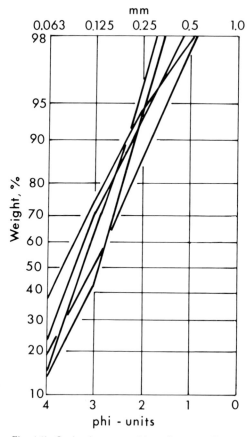

mm

Fig. 4.6b. Grain size composition of some sediment samples from sandbars and bottom of reservoir 1.

sediment actually measured at the outflow of the reservoir. During the rainy seasons 1973/74, 1974/75 and 1976/77, the reservoir was overflowing most of the time (Fig. 4.5). The minimum stages observed were in the period 1973/74 when, during long spells of rain-free days, the water temporarily could be stored within the reservoir pool despite some inflow to the reservoir. In 1974/75, the water level was never below the full supply level and the stages at the dam corresponded to the amount of water reaching the reservoir. This might explain the reduction in sediment accumulation in the reservoir between 1974 and 1975. Since 1976, almost all the water reaching the reservoir pool

diminishes as rapidly as in the rest of the river channel.

The measured sediment concentration in the water reaching and leaving the reservoir (Table 4.1 and Fig. 4.7) shows variations in the relationship between concentration at the inflow and outflow sections of the reservoir. This condition is partially due to the fact that the reservoir banks were covered with vegetation during some of the sampling periods, therefore reservoir deposition then might have been due to the effect of vegetation rather than to the size of the reservoir. In the rainy season 1973/74, some observations show lower sediment concentration in inflowing water as compared to water leaving the reservoir. No inflow sediment concentration analysis was made in 1976/77 and the concentrations at the outflow can be compared with the ones measured at the river station upstream of the reservoir. Nevertheless, this does not give the whole picture because several gullies empty into the stream between the station and the reservoir. The highest concentrations were observed in connection with heavy storms in January and March. The variations are very large between the observed concentrations at one and the same stage during the period of observation (Table 4.1).

The limited data presented here, and the changes in the shoreline of the reservoir between 1968 and 1975 (Fig. 4.4), show the transformation within a period of less than five years from a reservoir pool with maximum depth of 12 m to a low gradient channel reach with small pools alternating with riffles along the thalweg of the channel. Some of these pools were observed to migrate upstream in the period since 1975 (Fig. 4.4). Although the data provide an uncertain base for the estimation of sediment yield for the catchment, they give an example of rapid loss of water storage capacity by siltation in these reservoirs constructed by impoundment of the main streams and without any upstream provision for silt diver-

Table 4.3. *Notes on Relief, Soils, Vegetation, Erosion Features and Land Use: Roma Valley and Maliele catchments*

	Mountain Slopes	Basalt Foothills	Cave Sandstone Plateau	Escarpment and Scree Slopes	Lowland Pediments and Spur Slopes	River Valley Flats
RELIEF	200—500 m straight to concave slopes. 15°—45° gradients. Minor cliffs towards the top. Terraces of basalt lava flows.	0°—25° slopes and rounded to crested spurs separated by shallow streams.	0°—15° slopes; straight to concave in form grading into basalt foothills. Broad spurs separated by drainage depressions with relatively thick alluvial/colluvial fill.	Vertical cliff, 6—60 m high. Straight to slightly concave scree slopes with colluvial lobes and rock-fall debris.	0°—15° straight to slightly convex slopes, below scree slopes. Broad, rounded spurs separated by drainage depressions.	Level to almost flat, narrow plain along the main stream and its tributaries.
SOILS	Lithosols, gray to yellow loamy soils, shallow dark-surfaced soils in the lower parts of the slopes.	Thin, stony to thick dark-gray to black loams to fine sandy loams on open spurs. Dark-gray to black clays in the drainage depressions.	Bright reddish-brown to yellowish-brown loam and silty loams on the spurs, dark to very dark, mollic clays in the depressions.	Rocklands and colluvial materials grading to stratified, unsorted materials in the lower sections. These may be up to 10 m thick.	Reddish to yellowish brown, light gray to gray fine sandy soils; fine sandy loams 60—120 m thick.	Silt loams, to clay loams and clays. Dark, mollic and deep alluvial deposits with some thin layers of gravels.
VEGETATION	Indigeneous vegetation consists of *Themeda-Festuca* grassland. Small thickets of indigeneous trees and shrubs. Invasion of the lower sections by karroid bushes as a result of severe overgrazing.	Shrubs and indigeneous trees along drainage channels. *Themeda-Festuca* grassland.	Mainly grasslands modified by cultivation and overgrazing. The dominant grasses are of *Themeda-Cymbopogon-Eragrostis* transitional to Highland Sourveld.	Many of the south-facing slopes are covered with thickets of indigeneous trees and shrubs. The north to north-east facing slopes are covered by tall grasses (*Hyparrhenia* spp.). The original grasslands here are mostly protected from indiscriminate grazing, however, this has not stopped overgrazing.	*Themeda-Cymbopogon-Eragrostis* grasslands transitional to Highland sourveld, modified by cultivation and overgrazing. Severely overgrazed areas and abandoned fields are dominated by weeds (*Bromus* and *Poa* spp.) and karroid bushes (*Chrysocoma*).	*Themeda-Cymbopogon-Eragrostis* grassland. Willow trees (*Salix woodii*) and various types of reparian grasses (*Carex* spp. *Phragmites communis*, *Cyperas marginatus*, *Typha latifolia*).
EROSION FEATURES	Bare bedrock outcrops, sheet wash on fields, footpath rills, minor gullies along the streams.	Minor gullies. Sheet wash on spur-slopes. "cattle tracks", minor creep and regolith stripping.	Sheet wash and rills on cultivated lands and spur ridges. Splash/wind erosion indicated by bush pillars and pedestals. Bare ground between bush tussocks; gullies along main drainage channels. Regolith stripping at lower edges of the plateau. Local deposition.	Rockfalls, sliding and gullying. Minor rills in the lower sections of the scree slopes and along footpaths.	Gullies, sheet wash and rill formation on the fields, and open spur ridges. Deep gullies in the drainage depression. Piping in some areas. Local deposition in upper sections and in the drainage depressions.	River entrenchment, lateral bank erosion of main streams. Side gully-formation. Bar deposits and reservoir sedimentation.
LANDUSE	Pastures and limited cultivation.	Pastures, cropland and villages.	Cropland, pastures and villages.	Pastures and limited cultivation, woodlands.	Cropland, pastures, villages and reservoirs.	Cropland, pastures, tree plantations and reservoirs.

Table 4.4. *Trap Efficiency Estimates of Reservoir 1, using Churchill Curves, at Different Reservoir Capacities and Inflow Rates.*

Reservoir Capacity	Sedimentation index		% silt left in suspension at outlet to the reservoir	
m^3	^{q}mean	$^q50$	^{q}mean	$^q50$
3,472	3.0×10^4	3.0×10^5	90	50
2,234	1.2×10^4	1.3×10^5	99	60
1,830	8.3×10^3	8.4×10^4	100	70

sion. The study provides, however, clear evidence of high rate of delivery of sediment to the reservoir.

4.1.3 Catchment Surveys

The catchment surveys in this area included the following:

1. Field inventory of gullies, and the headward extension of some selected main gully heads.

2. Air-photo study of gully development between 1951 and 1971.

3. Recording of other erosion features in the catchment and the survey of one slope transect.

The gullies were all marked on the 1:10 000 air-photo maps including the different reaches showing recent erosion/deposition activity. The results of these surveys are summarized in Table 4.3 based on the division of the catchment into 5 landform groups.

The mountain and upper foothills regions are dominated by overgrazing and rills formed along livestock and human tracks between villages and along the main routes traversing the catchment. Very little gullying and surface wash are observable in the area. Rock outcrops dominate the mountain slopes, where grass vegetation is very scarce.

The rest of the highland region (Cave Sandstone plateau), is heavily cultivated and reveals three types of erosion: surface wash on open rounded spurs and spur slopes, minor rill formation in the fields on slopes steeper than 25%, gullying along the drainage depressions and regolith stripping at the lower edges of the plateau. Large areas of bare ground caused by overgrazing are found around homesteads.

The scree slopes show some sliding and several colluvial lobes with mixed materials of fine and coarse sediment up to large rock boulders. The lower edges of these slopes are entrenched by gullies continuing from the lowland region, especially in those areas where the slopes either form the valley-head slopes or extend to the river plain. Four fresh slides were observed in the area during the study period, and several old ones which have been colonized by vegetation. The slopes are mainly grazing lands. The high concentration of villages at the foot of the slopes has contributed to the extensive sheet and rill erosion in this zone caused mainly by livestock trampling and the high concentration of footpaths and roads leading from the villages.

The lowlands proper, lowland spurs and pediment slopes, are heavily cultivated and show severe surface wash and high concentration of gullies. These, however, form a very small fraction of the whole catchment.

Field measurements of gully extension in the area, together with air-photo comparison studies (Tables 4.5 and 4.3), reveal that the greatest headward extension of gullies is found in those gullies that are formed in the colluvial deposits. Very little bank erosion was observed in the area and most of the lowland river flats are marked by several zones of sedimentation in the form of point bars. In some reaches the main channel shows some downcutting into the older fluvial deposits.

4.1.4 Water and Sediment Discharge

As already mentioned in Section 2.2.2, there are no previous data on water discharge in the Roma Valley catchment.

Table 4.5. *Summary of Gully Extension within Roma Valley and Maliele Catchments Based on Air-photo Interpretation and Ground Measurements.*

Area	Gully head extension in metres 1951—1971								1973/74—1977						
	0—10	10—20	20—30	30—40	40—50	50—100	100	Total	0—2	2—4	4—6	6—8	8—10	10	Total
1	1		1	4				6		4	2				6
2		1		1		2		4	2	1	3	4		1	11
4									1	4	1				6
5												1			1
6.1			1		1	1		3	3	2	1			2	8
6.2		1					1	2	2	1			1	4	8
6.3		1		2				3	1	4	2			2	9
6.4					1	2		3	1	2	1	1			5
	1	2	2	8	2	3	3	21	10	18	9	7	1	9	54

Therefore two river stations were erected in 1973 for the observation of discharge and measurement of sediment concentration within the catchment. The upstreammost of these is at the outlet from Roma Valley area 1. The results of observations made at that station will be described in this section, whereas the results of the second station at the outlet of the whole of the Roma Valley catchment will be dealt with in Section 4.5.2.

The station is located upstream of a culvert on the Roma-Tloutle road and is referred to as St. Joseph's station in this report. The results dealt with are those of observations made between December 1976 and April 1977. Previous installations of the staff gauge were unsuccessful due to vandalism and lack of personnel.

The staff gauge was installed on 29 December 1976 and the cross-section surveyed four times during the rainy season 1976/77, at installation and on the three occasions of discharge rating. Relatively small changes were observed in the cross-sectional form during the study period and do not call for any adjustments of the rated discharges due to these changes.

The observed water stages at the station ranged from 40 cm (on the staff level), when all the flows were limited within the furrow on the left bank of the channel to 1.68 m when the culvert downstream of the section was overtopped.

The discharges observed during the period 29 December 1976 to 24 April 1977 are presented in Fig. 4.7, expressed in m³/day, and their frequency is shown in Fig. 4.8. The observed discharges, however, show an incomplete picture of the discharges during the period. Most of the peak floods are not included because they either occurred during the night or were so rapid that they were missed by the observers. However, the observed stages give a good picture of the position of the highest flows during the year, being concentrated in January and March. This should be compared with the rainfall data in Section 2.2.2. Even the rainfall data do not provide a complete picture because the rains responsible for discharges at the station often fall in the mountains without being recorded at the rainfall station, which is located downstream of the river station.

A mean discharge of 130×10^3 m³/day was obtained for the period (143 days), corresponding to a runoff of 5 mm/day from the catchment during this period. One significant feature of runoff in this area is its sensitivity to precipitation. This makes it as erratic as the precipitation both in time and space. The five months (December—April) used in this study cor-

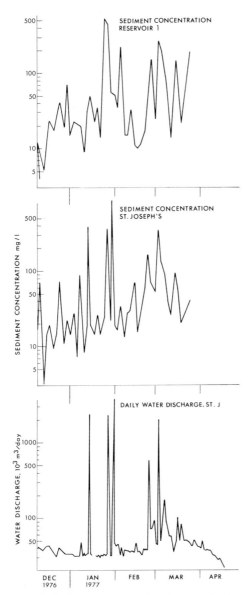

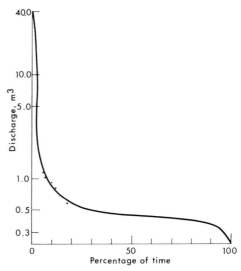

Fig. 4.8. Distribution of water discharges, at Liphiring river at St Joseph's river station. Dec. 1976—April 1977.

Fig. 4.7. Sediment concentration in water leaving reservoir 1, sediment concentration of water passing St Joseph's river station, and daily water discharge at St Joseph's river station during the rainy season 1976/77.

respond to the months of maximum rainfall in the area (Fig. 2.5) for the water year 1976/77 with another earlier rainfall peak in October. The precipitation in these months made up about 60% of the total rainfall for the water year. Due to the

fact that very little storage possibilities exist within the catchment, it could be expected that the discharge during the rainy season bears the same relationship to annual discharge as the monthly precipitation to the annual total precipitation. However, the influence on the river discharge of such factors as temperature, evaporation and use of water for domestic purposes, makes it very difficult to accurately extrapolate the runoff for the whole year from the short period of observation. The only certain conclusion that can be drawn from these studies is that a greater portion of the water reaching the catchment in the form of precipitation during the rainy season leaves the catchment in the form of short-lived floods following immediately after the occurrence of the rainstorms. During the period under study (December 1976—April 1977) the total discharge from the catchment was estimated to be 18600×10^3 m^3, corresponding to a runoff of 720×10^3 m^3 km^{-2}.

On 8 December 1976 a programme of water sampling was started at St. Joseph's station for analysis of sediment concentration. Up to the 29th the water stage was determined by measuring the water level below the lower part of the culvert block. Thereafter, the stages were read directly from a manual staff gauge installed upstream of the culvert. The sampling was done three times per week and on as many rainy days as possible. All in all 46 water samples were collected during the rainy period and sediment concentration determined (Fig. 4.7). Concentration figures were then transformed to suspended sediment discharge by multiplying the concentration by water discharge corresponding to the stage at the time of sampling. The sediment load in kg s^{-1} was then plotted against discharge on log × log paper using the 46 pairs. A least square regression line was calculated for the scatter of points (Fig. 4.9) and the line was found to be represented by the relationship.

$$L = 2.4 \times 10^{-7} \times Q^{1.8} \qquad (4.1)$$

where L is the suspended-sediment load in kg s^{-1}, Q the water discharge in m^3 s^{-1}.

This line is shown as a full line in Fig. 4.9. Visually, the scatter of points shows three trends: in the discharges below 0.5 m^3 s^{-1} between 0.5 m^3 s^{-1} and 26.0 m^3 s^{-1} and above 26.0 m^3 s^{-1}. Taking this into consideration, three lines were fitted by eye to the scatter of points and the mathematical expressions worked out for the lines which form a broken line curve represented by the group of equations:

$$L = \begin{cases} 2.0\ Q^{6.2} & \text{for } Q < 0.5 \\ 7.0 \times 10^{-2}\ Q^{1.4} & \text{,, } 0.5 < Q < 26.0 \\ 2.4 \times 10^{-3}\ Q^{2.7} & \text{,, } \quad Q > 26.0 \end{cases} (4.2)$$

The relationships expressed by these equations were used to compute sediment load for the rest of the observed discharges during the study period.

The large range of sediment concentrations for water discharges less than 0.5 m^3 s^{-1} show that the variation in sediment

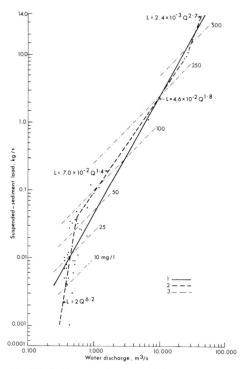

Fig. 4.9. Sediment rating curve, Liphiring river at St Joseph's. **1.** Regression curve based on all data, **2.** Line fitted to the scatter points by eye, **3.** Sediment concentration in mg/l.

concentration cannot be satisfactorily explained by variation in water discharge.

A total load of 5900 tonnes was obtained for the period 8 December 1976 to 15 April 1977, giving a mean daily suspended sediment discharge of 50 t/day for the period. Very little suspended sediment is transported during low flows of the winter months. Taking this into account and the fact that these months provide 60% of rainfall during the water year, would result in a sediment yield of about 380 t km^{-2} y^{-1}. This is probably an overestimation of the load, but it is of the right order of magnitude.

The figure for total load—5900 tonnes—is worth comparing with the estimate found for the sediment discharge at the outlet from the whole catchment during the same period. There, it was found that the total load for the period was 28 400

65

tonnes, giving an average of 200 t/day during the same period. These figures correspond to 4 t km^{-2}/day and 2 t km^{-2}/day for the downstream and upstream stations respectively. Thus there seems to be an increase in sediment delivery to the channel as one descends from the highlands to the lowlands. A glance at the measured sediment concentrations at the two stations confirms this observation.

This state of affairs should be expected from the physical and land use features in the area. The uplands are underlain by basalts and the soils formed there are less erodible as they tend to be more clayey than the soils formed in the sandstone terrain of the lowlands. Very little of the sandstone area is included in the upper catchment with outlet at the St. Joseph's station. The vegetation conditions are more inducive to erosion and entrainment of sediment by flowing water in the lowlands which is seriously affected by human activities such as intense cultivation of all available land and by settlements, especially the urbanization activities in the neighbourhood of the university and the Roma mission.

4.1.5 Results and Discussion

The reservoir rapidly filled between 1968 and 1972 and lost over 80% of its full supply volume. The siltation was rapid in the period December 1973 to December 1974 giving an annual rate of 30 cm/year. This rate was reduced to 13 cm/year in the period 1974 to 1975 when the last bottom surveying of the reservoir was made. Total sediment accumulation for the two year period was found to be 1642 m^3 (wet sediment). The mean grain size of the sediments accumulated in the reservoir (including sand banks above full supply level in 1973) is 0.15 mm with a silt-clay fraction making up 20% of the deposits. The dry matter was calculated to be 785 m^3 or 540 tonnes, corresponding to 15 m^3 km^{-2} y^{-1}, if it is assumed that the reservoir de-

posits form all the material eroded in the catchment above the reservoir. This assumption is false because in the period of study the reservoir trapped only a small fraction of the sediment reaching it, as can be seen from the sediment concentration of the water leaving the reservoir (Fig. 4.7), and the probable amount of material deposited in the reservoir by mean and median discharges of the stream feeding the reservoir during the rainy season of 1976/77 (Fig. 4.6). Using Churchill's trap efficiency estimation curves, the reservoir deposits were estimated to be 10–15% of the total load reaching the reservoir in 1973—74 and 1—40% in the period 1974—75, depending on the magnitude of summer flows. Thus the sediment yield for the catchment lies in the region of 350 t km^{-2} y^{-1}.

The results of the *catchment surveys* can be summarized briefly as follows. The area is severely eroded and erosion types and their intensities vary from landform to landform and land use category. Table 4.3 summarises these variations. The processes can be arranged as follows in order of their severity: sheet erosion and rill formation, gully erosion, landsliding and rockfalls, channel erosion and local deposition along the main channel. The most severe form of overgrazing is that found around villages in the form of bare ground areas, small rills and gullies formed along footpaths from villages to pastures and other villages. Gully erosion is more pronounced in the lowlands at the foot of the scree slopes down to the main stream's alluvial plain. The rate of gully growth seems to have decreased, as no new areas were invaded by gullies after 1951 and the greatest gully-head extensions occurred in the period 1951—61 with very small changes in the period up to 1977.

Water stage, discharge and sediment concentration measurements for the Roma Valley area 1, upstream of St. Joseph's culvert, cover the period December 1976 to April 1977. The reservoir stage at the outlet to

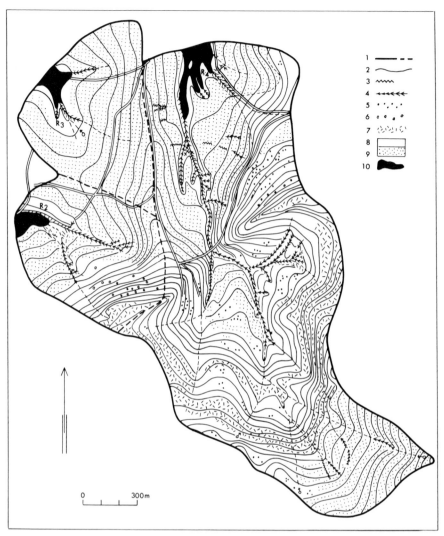

Fig. 4.10. Study areas 2, 3 & 4, Roma Valley. **1**. Catchment boundary, **2**. Contour at 5 m interval, **3**. Rills, **4**. Gullied reach, **5**. Homesteads, **6**. Scree boulders, **7**. Area of extensive bedrock outcrops and thin soils (rocklands), **8**. Pastures, **9**. Cropland, **10**. Reservoir.
Source: Thaba Bosiu Rural Development Project's ortho-photo maps (1:10,000, contoured), and air-photographs, taken in 1951/52, 1961/62 and 1971.

the catchment was measured during three periods: rainy seasons 1973/74, 1974/75 and 1976/77. These limited data form the basis of the estimation of stream flow and water discharge plus sediment load for the area. Daily values of discharge were calculated and found to range from 20×10^3 m^3 to 340×10^3 m^3 with a mean daily value of 130×10^3 m^3 for the period December 1976 to April 1977. This gives a

mean rainy season runoff of 58 l s^{-1} km^{-2}. The discharges equalled or exceeded 1%, 10%, 50%, 90% and 99% of the time were 3500, 78, 38, 26 and 21×10^3 m^3/day.

Sediment concentration of water passing the river station varied extremely and was low in December but increased later to a maximum value in February. The fluctuations of these values were related to type and intensity of rainfall producing the discharges and its areal distribution. At low flows the water was clear but the slightest rainfall occasion produced sudden turbidity in the water. The high concentrations in January and March can be attributed to high sediment production from cultivated lands at the time of ploughing and weed clearing.

The water discharge and sediment concentration values were used to compute sediment load for the season. A total of 5900 tonnes was obtained for the 129 days of observation, giving a mean daily sediment discharge of 1.6 t km^{-2}. The estimated sediment yield for the area is therefore 380 t km^{-2} y^{-1}. The sediment yield estimate obtained for the same catchment using reservoir accumulation and the trap efficiency estimate was 340 t km^{-2} y^{-1} (5—10% is trapped by the reservoir).

4.2 Roma Valley area 2

4.2.1 Introduction

The investigated area (Fig. 4.10) is a small catchment within the Roma Valley catchment (Fig. 2.1). The area of the catchment is 0.5 km^2 and has a relief ranging from 1670 m to 1870 m above m.s.l. (Fig. 4.11).

The major landforms in the area include plateau, escarpment, scree slope, pediment fans and topographic depressions filled with alluvial and colluvial sediments. The catchment is drained by a second order stream whose tributaries consist of very deeply incised gullies along the topographic depressions at the foot of the scree slope with a north-eastern aspect. The western channel flows on the west side of a minor alluvial/colluvial fan (Fig. 4.12) and has cut a deep gully into the deposits. The eastern gully consists of three distinct sections. The lowest and the uppermost sections are entrenched gullies and the mid-section is a wide, shallow, flat-bottomed gully and totally colonized by vegetation. Most of the material eroded upstream of this section is deposited in this reach. The result has been a thick deposition (max 3 m) which is being re-incised from downstream, just above a bedrock scarp.

Soils within the catchment are mainly loams and fine sandy loams (Rama and Tsiki series), stratified colluvial deposits characterised by coarse top layers, fine textured subsurface horizons, and the development of subsurface channels at contact zones between fine and coarse textured layers.

Land use and land management practices are the same as those in area 1 with the exception that no villages exist within this area.

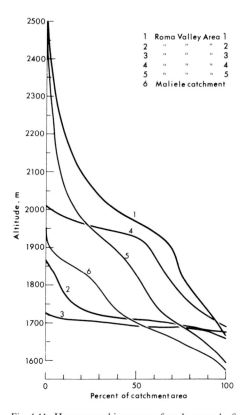

Fig. 4.11. Hypsopgraphic curves of study areas 1—6, Roma Valley and Maliele catchment.
Source: D.O.S. 421 (1:50,000, 1955—1966); Thaba Bosiu Rural Development ortho-photo maps.

Fig. 4.12. Ground photograph of the western gully and scree slope, Roma Valley area 2. Photo taken facing south (Photo: Q. K. Chakela, Nov. 1973).

The distribution of land use categories in the area is given in Table 4.6.

Table 4.6. *Land Use Categories in Area 2, Roma Valley Catchment.*

Land Use Category	Percent of total area
Cropland	53
Gullied area	8
Rockland	5
Open pasture	34

4.2.2 Reservoir Surveys

The reservoir at the outlet to Roma valley Area 2 is a chimney-spillway type. The construction was completed in 1968, thus when these studies were begun the reservoir had had five years of sedimentation life. The reservoir depth and volume at completion is not known and it has not been possible to reconstruct the original

data as the reservoir has been deepened and widened by excavation of the original gully across which the dam was built.

The first survey of the reservoir was made in 1973 and a map was produced showing the bottom topography of the reservoir. The soundings were repeated in 1975 and 1977. These data were used to produce three bottom maps of the reservoir (Figs. 4.13a—c). The area at different levels (in m below full supply level) and volumes of the reservoir were computed from the three maps and capacity curves were drawn for each survey (Fig. 4.14). The total volume decrease between 1973 and 1977 was taken to be the amount of sediment accumulated in the reservoir and was found to be 4405 m^3, composed of 3818 m^3 and 587 m^3 for the periods 1973—75 and 1975—77 respectively.

Some core samples were collected from the bottom of the reservoir in 1975 and 1977 during the bottom surveys. The 1975

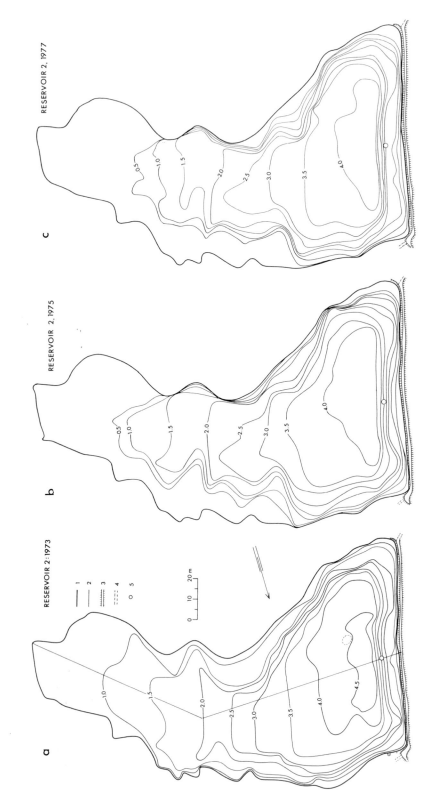

Fig. 4.13. Bottom maps of reservoir 2. a) October 1973, b) December 1975, c) April 1977. **1**. Full supply level, **2**. Contour line in m below full supply level. **3**. Reservoir embankment, **4**. Road, **5**. Chimney pipe spillway.

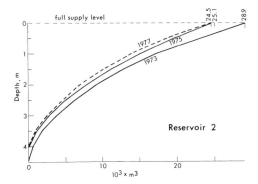

Fig. 4.14. Capacity curves, reservoir 2, 1973, 1975 and 1977. 28.9, 25.1 and 24.5 are full supply capacities of the reservoir in 10^3 m^3 for the indicated years.

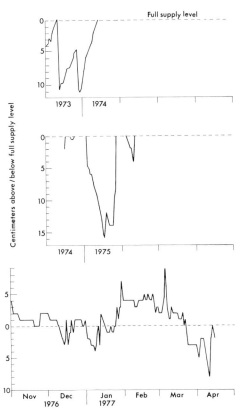

Fig. 4.15. Water level records in reservoir 2 during the rainy seasons, 1973/74, 1974/75 and 1976/77. The staff gauges for 1973/74 and 1974/75 recorded only from full supply level and less.

samples were not weighed in the wet state but transferred to glass beakers and weighed after drying in an oven at 105°C overnight. The 1977 samples were weighed in the field with a spring balance and wet bulk density determined.

The water level in relation to full supply level has been recorded since December 1973 but with gaps during the periods the author was away from Lesotho. The record is thus made up of three periods in the rainy seasons 1973/74, 1974/75 and 1976/77 (Fig. 4.15).

4.2.3 Catchment Surveys

The catchment surveys in this catchment consisted of three types of data collection: field inventory of gullies and their headward extension between 1973 and 1977; air-photo study of gully development between 1951 and 1971 based on air-photographs taken in 1951, 1961 and 1971; and slope surveys along the two main channels feeding the reservoir and across the main side slope.

These data are summarized in Table 4.5, Figs. 4.16 and 4.17. In connection with slope surveys an attempt was made to measure the length of bare ground along the transect and convert this to percentage grass cover. The eastern channel gully head was very active during the study period and it was deemed necessary to produce a map of the actively eroding section and to make a rough calculation of the volume changes of the gully.

4.2.4 Results and Discussion

The catchment is dominated by cultivation. Most of the fields within the catchment show high activity of surface wash and formation of rills. The clearest sign of wash is soil accumulation on the upslope side of the buffer strips and field boundaries, exposure of pale-coloured subsoils with very little organic matter, and formation of rills on the downslope side of the boundaries and buffer strips.

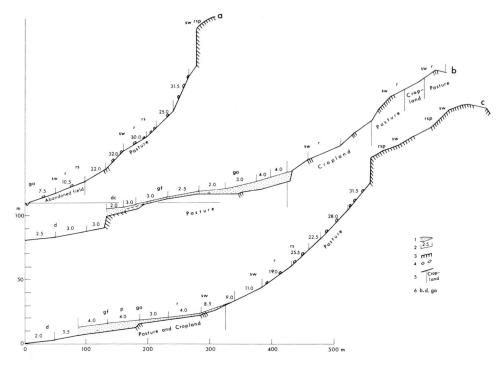

Fig. 4.16. Slope profiles and longitudinal profiles of major gullies in Roma Valley Area 2. Ground measurements extended to below the vertical cliff for a, c and to the first bedrock outcrop for b. The rest of the profile is extrapolated from ortho-photo maps with 5 m contour interval. **1**. Surface profile, **2**. Gully bottom profile, **3**. Rock outcrop, **4**. Slope gradient in degrees, **5**. Boulders—P = Pastures, C = Cropland, d = deposition, dc = entrenchment of gully fill, ga = actively eroding gully reach, gf = gully bottom deposition, r = rill erosion, sw = surface wash, rs = rock screes, rsp = regolith stripping.

The area along the gullies and on the main scree slope is severely overgrazed. Overgrazing manifests itself in three forms: areas of short grass stands with complete cover, especially in the low-lying area dominated by short grasses (*Eragrostis, Themeda* and *Cynodon* spp.); areas dominated by bare ground patches between grass tussocks, largely at the foot of the scree slopes where the main vegetation is tall grasses like thatching grass (*Andropgon* sp.); and areas invaded by Karroid bushes (*Chrysocoma* sp.) and bitter grasses (*Enionurus argenteus*).

The third erosion damage in the area is due to gullying. Two main gullies and their short side gullies dominated the area. There seems to have been stabilization of the lower section of the gullied reach. This may partly be an effect of the changed flow conditions upstream of the reservoir, but it may also be a natural aggradation in a reach with a reduced slope gradient.

From air-photo comparison, no new gullies seem to have formed since the 1950s. The headward extension of the main gullies occurred mainly between 1951 and 1961 whereby the western gully extended 150 m and the eastern gully extended 50 m, producing the presently active reach (Fig. 4.17). The western gully has reached the bedrock and a very thin soil cover area, thus it has no chance of further headward extension. However, the lowest parts of the entrenched reach are eroding laterally, along contour furrows, through pipe collapse and slumping. The minor side gullies of this gully show very slow rates of growth.

The left gully is characterised by three

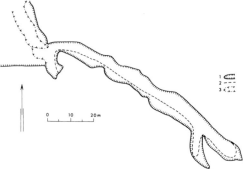

Fig. 4.17. The uppermost actively eroding section of the eastern gully, Roma Valley area 2.
a) Ground photograph (Photo: Q. K. Chakela, April, 1977)
b) Planimetric Map (March, 1977). **1**. 1977 gully limits, **2**. 1973 gully limits, **3**. Secondary scarps within the 1977 gully walls.

reaches dominated by deposition or erosion. The section below the sandstone scarp upstream of the reservoir is presently stable and shows very little activity, it has a floor completely colonized by grass, making it a depositional zone continuing to the reservoir. Just above the bedrock scarp, the gully is wide, flat-bottomed and has very shallow side walls. The bottom material consists of laminated deposits of thin, alternating layers of sand and clay-silt materials. The reach is completely overgrown with grass and becomes swampy in the lower sections during the rainy period. There is fresh entrenchment and stripping off of the deposits above the bedrock scarp. Two deeper gullies have been formed on either side of the main channel close to the side banks. Between 1973 and 1977 these two gullies showed a headward extension of 5 and 3 metres respectively. The third section in this gully consists of the actively eroding gully head (Fig. 4.17) which increased in area and depth between 1973 and 1977 by 83 m^2 and 0.7 m respectively. Most of the

73

blocks in the gully have been stabilized in position by vegetation. Most of the material from this gully section is deposited in the grassed reach below and very little material reaches the reservoir.

The results of reservoir surveys give a total sediment accumulation of 4405 m^3 between 1973 and 1977. The mean wet bulk density of the deposits determined through core sampling was found to be 1.6 g cm^{-3}. Assuming particle density of 2.65 g cm^{-3} for the materials deposited, the accumulated dry volume of the material is 2775 m^3. Two values of dry bulk density were computed for the materials. One from the wet bulk density measured in the field and dry bulk density determined after oven drying of the core samples. The mean values obtained were 1.03 g cm^{-3} and 0.95 g cm^{-3} respectively. Using these two values the mean annual sediment accumulation between 1973 and 1977 is found to be 880 t y^{-1} and 800 t y^{-1} respectively. These would correspond to annual sediment yields of 1870 t km^{-2} and 1700 t km^{-2}.

The volume of the deposits from the gully activity was estimated to be 600 m^3. Most of these deposits never reach the reservoir. The blocks are either stabilized within the gullies or deposited in the level reaches of the gullies where vegetation acts as an effective trap for the material. From these observations it can be concluded that the bulk of sediments in the reservoir are a result of wash and rill erosion with little gullying.

The erosion and sedimentation processes within Roma Valley area 2 can be summarized as follows: gully erosion is the most obvious erosion damage in the area. However, since 1950 no new gullies have appeared within the catchment. The gullies grew rather rapidly in the initial stages, but are now very slow in both headward and lateral growth. This is mainly due to three factors. First, the main gullies have grown beyond the point of headward extension and therefore continue to the water divide as small shallow rills with bedrock bottom. Secondly, the eastern gully's mid-section has reached a semi-balance between width and depth and is aggrading and colonized by vegetation which tends to increase the trap efficiency of the reach. Thirdly, the section of the entrenched reach downstream of the confluence has been affected by the reservoir and the fact that it has widened to a point where linear flows are minimum and inducive to sedimentation.

Sedimentation within the catchment occurs in two zones. In the low-gradient reaches, within the wide bottom part of the eastern gully and just above the reservoir; and within the reservoir. The gully section downstream of the reservoir is stabilized just below the reservoir and colonized by vegetation. This is due to the fact that water rarely flows out of the reservoir and when it does, is robbed of its energy by the chimney-spillway fall and reaches the lower part of the reservoir with very little erosive power. This, coupled with the high sediment concentration of the outflows, leads to sediment deposition which is later colonized and stabilized in position by vegetation.

Rill erosion within the catchment is limited to the cultivated areas and recently abandoned fields. Here it is accompanied by severe surface wash during parts of the rainy season, during ploughing and just before the crops and weeds on the fields and fallow have established themselves firmly enough to check the removal of soil after thunderstorms. The fact that the fields are opened for grazing during the dry season helps to loosen the soil through livestock trampling.

The scree slope is covered with boulder lobes which witness an earlier rockfall activity from the escarpment cliff. Within these zones the greatest damage is due to overgrazing and burning of the grass and consequently increased surface wash.

The major processes in the area are surface wash and raindrop splash, and gully erosion in that order of magnitude.

The effect of contour and buffer strips is minimal and some of the contour furrows have been the location for the initiation of side gullies. Gully head stone walls seem to suffer the same fate. The gullies eroded below the stone constructions along bedrock/regolith interface by means of seepage of water and formation of pipes. The most effective anti-erosion measure in the area has been the reservoir. However, the effect is limited to a very small area upstream and downstream of the reservoir where there is positive evidence that the aggradation taking place is a result of the reservoir.

4.3   Roma Valley area 3

4.3.1 Introduction

This is a small catchment (0.6 km²) formed in a bedrock depression of Red Beds formation and crossed by two narrow dolerite dykes which form no prominent landform element within the area. The topographic depression in which the catchment is located is formed on one of the larger pediment slopes within the Roma Valley catchment described in Chapter 2 (Fig. 2.1). The catchment lies between study areas 2 and 4 (Fig. 4.10) and has a relief of 50 m (1675—1725 m above m.s.l.) and general slope gradient of 6%. The central part of the catchment is occupied by a system of gullies consisting of two major branches. At the head of the western gully are minor swamps with intermittent springs.

The catchment is dominated by three soil series:

1) Dark reddish brown, sandy loam to clay loam with moderate to good drainage, rapid runoff and medium permeability in surface layers and decreasing with depth. It occurs along the eastern and southern boundaries of the catchment. This soil series is called Leribe loam.

2) Dark brown, sandy loams with reddish brown B-horizon, poor drainage, medium runoff which may be moderately high in subsurface layers (Rama loam). It has moderate permeability which might be high in individual horizons above clayey or lithic contact. This makes the soil series liable to formation of subsurface pipes.

3) The third soil series (Tsiki fine sandy loam) has a dark brown light loam in surface layers with blocky structure and friable consistence. The deeper horizons have heavy loam texture and prismatic structure to platty structure, light brownish grey and brown colours. Where not drained by gullies, the soil series is poorly drained. It has low permeability in the subsoil leading to lateral movement of water in or on the surface soil. The surface runoff is high.

Land use within the catchment can be divided into three categories:

1. Grazing lands along the gullied area and between the fields in the summer months and grazing of the crop stub after harvest.

2. Cropland, used for growing staple crops like maize, beans and sorghum.

3. Fish culture in the reservoir and reservoir-fed ponds below.

The land management practices in this catchment are similar to those of catchment 2, but there is no irrigation practice here. The land use categories can be summarized as follows: 84% cropland, 4% gullied area and reservoir, and 12% open pasture.

4.3.2 Reservoir Surveys

Roma Valley area 3 reservoir is a chimney-spillway type. The dam wall is 175 m long and is made of soil material excavated from the reservoir. The wall is planted with kikuyu grass. The dam was completed in 1968. There is no data as to the original capacity immediately after completion (Fig. 4.18a & b).

Fig. 4.18a. View over Roma Valley area 3, its reservoirs and fish ponds. Reservoir 3 in the foreground. Note the intensity of cultivation. The cattle, just above the right lower corner, are grazing on already harvested field. Profile 1 on Fig. 4.19 measured from the second inflow channel from the left, along the gully, and then continuous to the right of and beyond the cattle. (Photo: Q. K. Chakela, April 1977.)

Fig. 4.18b. Roma Valley area 3. Reservoir 3 in the centre of the photograph and Roma Mission in the background. Note the actively eroding gully to the right along which the upper profile on Fig. 4.19 is measured to just beyond the photo point. The black-coloured areas show recent ploughing. (Photo: Q. K. Chakela, Nov. 1973.)

The first survey was done in December 1973 and consisted of bottom sounding and mapping of the full supply level shore line. The full supply level is taken as the level of the top of the spillway drum. In 1975 the first transversal profiles were resurveyed. The final survey was made in April 1977. The data from 1973 and 1977 were used to produce two bottom maps (Figs. 4.20a & b). The volumes of the reservoir were computed from these maps and capacity curves drawn (Fig. 4.22).

The water level of the reservoir in relation to full supply level has been recorded since 1973. The record is not complete and consists mainly of observations made during the rainy seasons of 1973/74, 1974/75 and 1976/77 (Fig. 4.21).

Some water samples were taken for sediment concentration analysis at the two inflow channels in 1973 and 1975 and near the spillway. The spillway was never overtopped during the whole time the observations were made and thus it has been impossible to estimate the concentration of the water leaving the reservoir. The main contributing factor to this fact is that the reservoir's gated outlet was opened occasionally during the rainy and dry seasons to supply water to the fish ponds immediately downstream of the reservoir.

4.3.3 Catchment Surveys

The catchment surveys in this area consisted of air-photo comparison of the gully sections upstream of the reservoir and a general survey of the erosion conditions in the area, using photographs from 1951 and 1961. The gullies showed no change in form or length between the two dates nor was any change visible in comparison with later photographs. The second set of observations dealt with the annual measurement of the northern inflow gully which is the only one showing any possible head-

ward extension between 1973 and 1977. Due to the fact that the area is mainly cropland, sheet erosion and rill formation studies could not be made from air-photo interpretation. The field studies in 1973/74, 1974/75 and 1976/77 revealed that wash and rill erosion was very active within the cultivated lands. This could be deduced from small rills that form across the ploughlines during the ploughing time after rainstorms, large sediment accumulation upstream of buffer strips and field boundaries, and small rills and exposure of subsoils downstream of the buffer strips. The area most severely affected by these two types of erosion is upstream of the northern inflow channel and along the side slopes to the west of the catchment.

Three slope transects were measured along the main inflow channels up to the water divide, and during these measurements observations were made in the immediate neighbourhood of the transects of erosion processes, gully depth and form, and amount of grass cover. Two of these are given in Fig. 4.19.

4.3.4 Results and Discussion

The reservoir surveys within Roma Valley area 3 (Fig. 4.20) revealed that the reservoir had changed in volume slightly between 1973 and 1977, and that the change

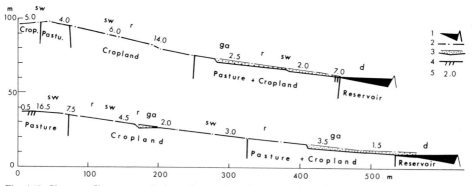

Fig. 4.19. Slope profiles measured along the northern (upper) and southern (lower) main feeding stream channels of reservoir 3. **1**. Reservoir, **2**. Measured segments, **3**. Gullied reach, **4**. Rock outcrops, **5**. Slope gradient in degrees; d, ga, r, sw dominant processes explained in Fig. 4.16.

77

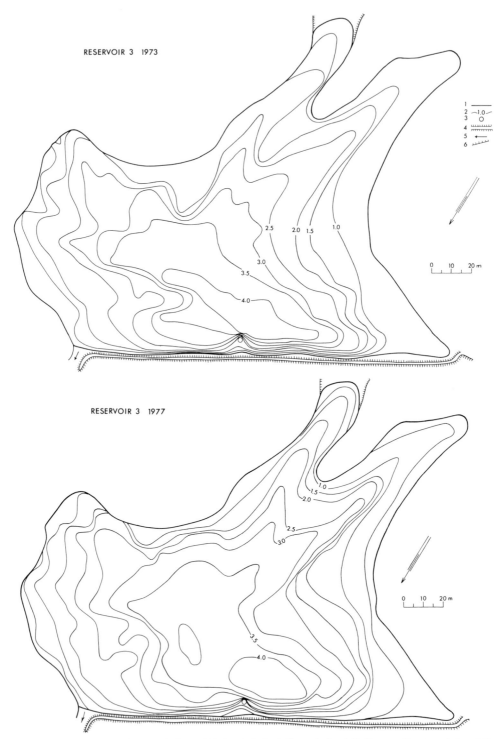

RESERVOIR 3 1973

RESERVOIR 3 1977

Fig. 4.20. Bottom maps of reservoir 3. a. 1973, b. 1977. **1**. Full supply level, **2**. Contour line in m below full supply level, **3**. Chimney pipe spillway, **4**. Reservoir embankment, **5**. Overflow spillway, **6**. Gully (un-surveyed).

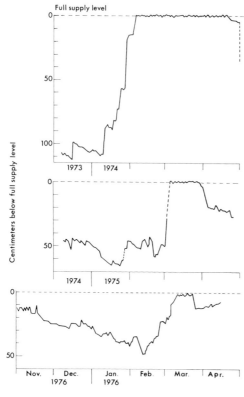

Full supply level

Centimeters below full supply level

1973 | 1974

1974 | 1975

Nov. | Dec. | Jan. | Feb. | Mar. | Apr.
1976 | 1976

Fig. 4.21. Water level in reservoir 3 during the rainy seasons 1973/74, 1974/75 and 1976/77.

was an increase of the reservoir volume by 270 m³. This is contrary to what could have been expected. The capacity curves show that only in the deepest parts of the reservoir has there been an aggradation (Fig. 4.22). Two explanations can be given for this observation. First, the gully system across which the reservoir dam is constructed is formed in sediments liable to formation of subsurface tunnels. Thus, collapse of the roofs of these tunnels in the intervening period would make the reservoir appear to have larger volume than in 1973 as the volume of the materials from the pipe roofs is very small compared to the pipes themselves and would contribute very little to the sediment within the reservoir. Secondly, the reservoir was flushed out several times during the study period to supply water to the fish ponds. This leads to removal of some or all the material deposited in the reservoir in the immediate neighbourhood of the gates, and also to the erosion of the gully banks of the submerged gully system due the repeated fluctuation of the water level. In this way the reservoir capacity is maintained at the origin value, or increased.

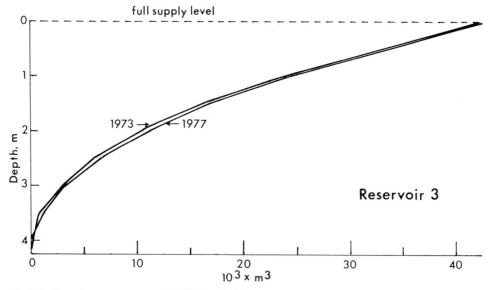

full supply level

Depth, m

1973 → ← 1977

Reservoir 3

$10^3 \times m^3$

Fig. 4.22. Capacity curves, reservoir 3, 1973 and 1977.

The deepest part of the reservoir decreased in depth, and this decrease is below the gated outlet (3.0 m below the chimney spillway top). The observed reservoir aggradation is so little that the reservoir studies in this case supply no conclusive information on the rate of erosion in the area.

The air-photo comparison of the gully conditions between 1951 and 1961 showed no significant changes in the shape, size or length. The present gully-heads, now partly submerged in the reservoir, are below a bedrock scarp and the gully system below the reservoir has a bedrock bottom. The bedrock scarp slowed down the head-retreat of the main gully heads. The bedrock bottom impeded the down-cutting and widening was limited by the fact that the gully width had become so large that concentrated flow was impossible. The rest of the catchment showed very little change between the two photography dates. This is mostly due to the fact that 84% of the area is cropland and the only evidence of erosion damage in the area are the "terraces" formed along the buffer strips and paler gray tones of subsoil downstream of the buffer strips caused by removal of the organic layers of the soils.

The field studies in the rainy seasons of 1973, 1974/75, 1975/76 and 1976/77, however, revealed that overland flow and rill formation were actively removing the soil from the croplands during the early summer before the weeds and crops have established effective surface cover to impede raindrop impact. This is evidenced by the small rills and sediment accumulations which appear on the fields after rainstorms of high intensities, and small alluvial fans formed upstream of buffer strips at the mouths of small rills, some of which run the whole width of the interbuffer strip zone, depending on the slope and the width of the zone.

The gullies are not very active but one gully-head (Fig. 4.19 top) showed an extension of 7.6 m between 1974 and 1977,

and a total volume increase of 84 m³. The other channel has a discontinuous gully system with two entrenched reaches separated by an aggrading zone (Fig. 4.19 lower).

The reservoir surveys in this area gave no conclusive information on the rate of erosion and sedimentation. The catchment surveys, however, showed that there is a very severe surface wash and rill erosion taking place in the area, but it was not possible to quantify this due to the land use pattern here. The materials transported along the rills and as wash load rarely leave the cropland area and the main portion is arrested by the buffer strips. The gullies have a limited activity and seem to play a very minor role in the environmental change except for draining the marshy areas upstream of the reservoir and as main transport channels for the material eroded in the upstream reaches. The important observation which can be made from these studies is the effect of reservoir emptying on the rate of the silting of the reservoir. This may have led to the fact that after almost ten years the reservoir still has its full capacity, although this might have taken place at the risk of some further erosion in the upstream part of the dam.

4.4 Roma Valley area 4

4.4.1 Introduction

Roma Valley area 4 is the second largest of the subcatchments studied within the Roma Valley catchment (Fig. 2.1, Chapter 2).

It has an area of 2.2 km² and ranges in altitude from 1695 m at the reservoir outlet to 2000 m above m.s.l. (Fig. 4.11). Two major dolerite dykes cross the catchment. The first of these (Fig. 4.23a) divides the catchment into two parts just below the Cave Sandstone escarpment. The major landforms in the area are a basalt/sand-

Fig. 4.23a. Roma Valley area 4. Dolerite dyke dividing the catchment into two. (Photo: Q. K. Chakela, April 1977.)

Fig. 4.23b. View over the lower reaches of Roma Valley area 4 and the location of reservoir 4. Note the grass covered gully slopes to the left of the main stream channel and the dense cover of wattle trees along the banks of the main stream channel. (Photo: Q. K. Chakela, Feb. 1975.)

stone plateau above the 1900 m contour, an escarpment, scree slopes, spur and pediment slopes and a narrow alluvial plain along the main channel.

The major soil series in the catchment are Matšaba-Ralebese loams, Popa-Matšana loams and rock outcrops in the basalt terrain, colluvial materials on the scree slopes, and Tsiki fine sandy loams, Leribe loams and Rama loams in the lowlands. The first two series occupy the ridges of the lowland spurs, and Rama loams occupy the central part of the catchment. Along the gullied reach near and upstream of the reservoir, wattle trees have been planted and the main stream is flanked by

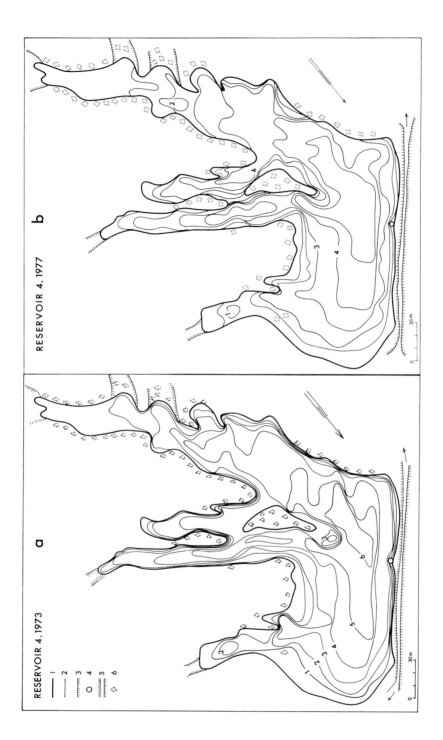

Fig. 4.24. Bottom maps of reservoir 4. a) Nov., 1973, b) March, 1977. **1**. Original full supply level, **2**. Contour lines in m below full supply level, **3**. Unsurveyed gully, **4**. Chimney pipe spillway, **5**. Reservoir embankment (earthen), **6**. Trees. Note the position of 1977 full supply level (o contour line) at the mouths of the feeding stream channels.

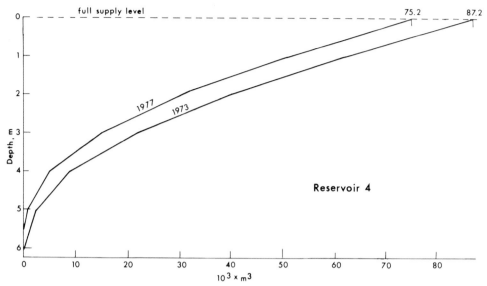

Fig. 4.25. Capacity curves, reservoir 4, Nov. 1973 and April 1977. The figures 87.2 and 75.2 are full supply capacities in 10^3 m^3 for 1973 respectively 1977.

willow trees here and there (Fig. 4.23b). Some poplar groves have been planted in some areas around villages.

Land use in this area is the same as for areas 1—3 above. The different land use categories have the following distribution in the area: cropland 36%, gullied area and reservoir 6%, rockland 16%, pastures and villages 42%. The boundary between rockland, pastures and villages is very diffuse. In the category gullied area, the minor gullies on cropland and pasture areas are not included.

The fields in the immediate neighbourhood of the reservoir were irrigated in the period 1970—75 but this had stopped in 1977. Fish have been planted in the main reservoir and in small fish ponds downstream of the reservoir. The other land management practices are the same as those applied in areas 1—3.

4.4.2 Reservoir Surveys

The reservoir in Roma Valley area 4 is an earthen wall dam, 175 m long and built across a gully system with two main gullies separated by two smaller ones. The spill-

way is a chimney pipe concrete drum with a depth of 6.3 m.

The reservoir was surveyed in November 1973, partly in March 1975 (only the first 8 transverse profiles) and finally in April 1977. The depth measurements were used to produce two bottom maps of the reservoir (Fig. 4.24). Total volumes of the reservoir were calculated for the two measuring dates. The volumes enclosed by different contours and the depth were used to construct capacity curves of the reservoir in 1973 and 1977 (Fig. 4.25), and the volume difference between 1973 and 1977 was calculated.

A manual staff gauge was installed near the spillway drum in 1973 and daily readings of water level, in reference to full supply level, were taken. A programme of water sampling of water leaving and entering the reservoir was also started. The water samples were taken during rainy days and three times a week during the spill-over period. However, these records are limited to the time the author was in Lesotho (Fig. 4.26).

In 1973 two small sandbanks had begun to develop at the mouth of the main inflow

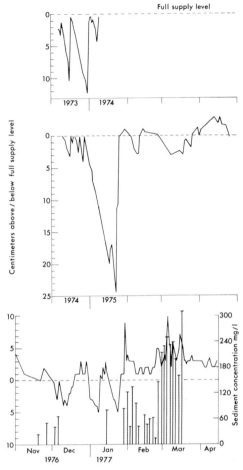

Fig. 4.26. Water level records during the rainy seasons 1973/74, 1974/75 and 1976/77, and sediment concentration (T-bars) at the outflow section during the rainy season 1976/77, reservoir 4.

channel to the reservoir. Some sediment samples were taken from these and analysed for grain size distribution (Fig. 4. 27a).

4.4.3 Catchment Surveys

Catchment surveys in this area were not as extensive as for the other areas. They included an air-photo interpretation of erosion conditions in the catchment from 1951, 1961 and 1971 air-photographs. However, no quantitative measurement of gully growth was made. This was mainly due

to the fact that the trees planted along the main gully between 1951 and 1961 made it difficult to observe changes in gully sides. Besides air-photo interpretation, ground observations of gully-head extension of 21 gully-heads were started in 1973 and annual observations of channel and gully lateral extension were made along the major gullies (Table 4.12). Two of the gullies were mapped (Fig. 4.28). A slope transect was measured along the eastern gully to the pediment slope above it. These measurements included gully depth, gradient and vegetation cover beside the gully.

4.4.4 Results and Discussion

The dominant erosion types within Roma Valley area 4 are wash-, rill- and gully erosion, regolith stripping and furrows caused by footpaths. Wash erosion is dominant on cultivated and pasture lands on spur slopes. It manifests itself in the form of exposure of pale-coloured subsoils, bush pedestals and sediment accumulations upstream of buffer strips and minor alluvial fans at the foot of spur slopes. Gullies are of two main types: continuous gullies forming headward extension of major streams and their tributaries, discontinuous gullies formed on spur slopes and ending at the break of the slope to the alluvial plain of the main channel. To these two main types should be added rills and gullies formed along the footpaths. In the foothills region the dominant erosion types are stripping of the regolith just above the Cave Sandstone, rill formation and surface wash on cultivated lands, bare ground areas around homesteads caused by livestock, and gullies along the drainage depressions. The scree slopes show very little erosion but severe overgrazing.

The ground measurement made between 1973 and 1977 of the gully headward extension shows that gully growth varied from 1 m to 6 m in the four years of observation. This growth is not continuous and

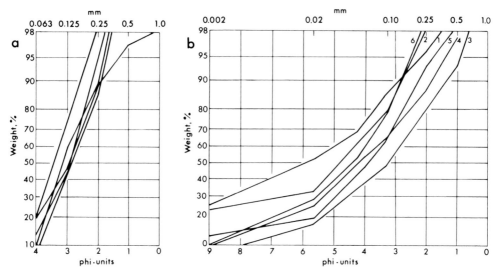

Fig. 4.27. Grain size composition of a) Sediment samples from the sides and bottom deposits of reservoir 4, b) A-horizons of soil series of the area. **1.** Fusi loam, **2.** Leribe loam. **3.** Matšana loam, **4.** Popa loam, **5.** Rama loam, **6.** Tsiki fine sandy loam.

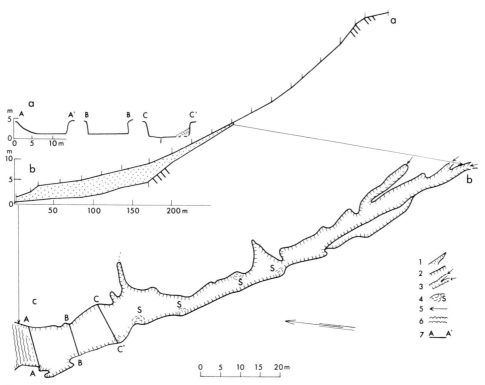

Fig. 4.28. Transverse profiles, longitudinal profile and planimetric map of the eastern inflow channel of reservoir 4. The longitudinal profile is continued beyond the gully-head scarp to sandstone terrace upstream. **1.** Gullied reach, **2.** Bedrock outcrops, **3.** Side walls of the main gully, **4.** Slumped zones, See also transverse profile C—C′ right bank, **5.** Major inflow routes into the gully, **6.** Reservoir 4, **7.** Location of the transversal profiles.

several gullies included in the observation net have not shown any headward growth.

The two main gullies ending at the reservoir show localized lateral extension in the form of block slumps (Fig. 4.28a & b). The blocks seldom leave the gully area but are stabilized in position by vegetation. This leads to decreases in gully depth and increases in width. The trees planted along the main stream banks have minimized lateral erosion and where this occurs the slumped blocks are held in the neighbourhood of the scars by tree roots. The major conclusion from these observations is that the materials eroded in the main gullies seem to travel very short distances and constitute a minor portion of the reservoir deposits.

The median grain size of the A horizons of the main soil series in the catchment is in the range 0.02—0.11 mm and their clay-silt fractions (finer than 0.063 mm) make up 38—70% of the grain size distribution. The corresponding values for sandbank deposits are 0.12 mm and 7—20% (Fig. 4.27).

The reservoir volume was reduced in capacity by 14% between 1973 and 1977. The reduction is in the form of aggradation at the mouths of the two main feeding streams, reduction in depth from a mean depth of 3.2 m in November 1973 to 2.90 m in April 1977 and reduction in maximum depth of the reservoir from 6.2 in 1973 to 5.80 m in 1977. The total decrease in capacity between 1973 and 1977 is 12100 m^3 and is illustrated by the capacity curves (Fig. 4.25) which show that the greatest changes took place in the shallow parts of the reservoir.

The mean wet bulk density of the reservoir sediments was found to be 1.53 g cm^{-3}. If the mean particle density is taken to be 2.62 g cm^{-3} (mean particle density of sediments taken from reservoirs in areas 1, 2 and 6), a dry bulk density of 0.88 g cm^{-3} and a total dry weight of 6000 tonnes are obtained for sediment accumulated between 1973 and 1977. This value corre-sponds to a sediment yield of 800 t km^{-2} y^{-1} for the catchment. The rate of siltation of the reservoir is found to be 25 cm/year (6.5 g cm^{-2} y^{-1}).

The amount of sediment accumulated in the reservoir represents only a fraction of the total sediment delivered to the reservoir. The other portion is carried out with the spill-over water to the lower reach in the form of suspended sediment. In estimating the percentage of the total sediment reaching the reservoir made up by the material accumulated in the reservoir, several factors have to be taken into consideration. Among these are the period of water retention in the reservoir, the duration and amount of spill-over into the lower reach, the characteristics and amount of sediment load of the river feeding the reservoir, depth and volume of the reservoir, and the location of the outlet. These factors determine the relative sediment retaining capacity of the reservoir, known as trap efficiency (Fig. 4.29).

As a first approximation of the spill-over load, sediment concentration of water leaving the reservoir during spill-over can be used. Before the full supply level of the reservoir is reached, no sediment is lost. Measurements of sediment concentration of water leaving the reservoir were made in the 1973/74 and 1976/77 rainy seasons (Fig. 4.26) and the water levels in relation to full supply level were recorded for the three rainy season 1973/74, 1974/75 and 1976/77. A mean concentration of 120 mg/l was obtained during the 1976/77 rainy season (range 40—350 mg/l). The mean rate of outflow for the period November 1976 to April 1977, was estimated to be 10 l/s, with a minimum outflow of 400 l/s recorded after a rain storm in March. These two values give a mean sediment discharge of 100 kg/day for the rainy season. The duration of spill-over for the three seasons, 1973/74, 1974/75 and 1976/77 was 317 days (Fig. 4.26). The estimated duration of spill-over time in the 1975/76 rainy season is 67 days, thus for the dur-

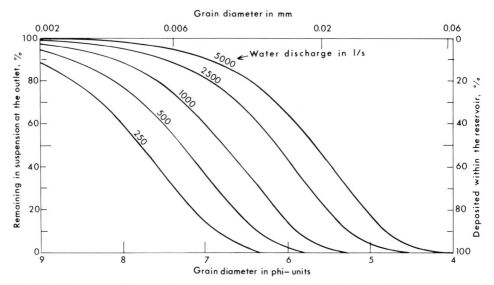

Fig. 4.29. Relative trap efficiency of reservoir 4 based on Sundborg formula. The calculations are based on water temperature of 20°C, sediment particle density of 2.65 g cm⁻³, and that on the average 30% of the reservoir volume took part in throughflow.

ation of the surveys the spill-over time is 384 days. The spill-over load is then 38.4 tonnes, bringing the total sediment yield to 825 t km⁻² y⁻¹. The spill-over period is so short that the original figure can be taken as valid.

In conclusion, the dominant erosion types in the area are surface wash erosion, gully erosion, rill erosion and bedrock stripping. The sedimentation takes place in the area in gully bottoms, in small alluvial fans at the foot of the spur slopes, at the mouths of the channels feeding the reservoir and in the reservoir. The area has a sediment yield of 900 m³ km⁻² y⁻¹ (800 t km⁻² y⁻¹) and the rate of reservoir sedimentation is 25 cm/year (7 g cm⁻² y⁻¹). These figures are reasonable estimates of the rate of siltation and erosion in the area considering that the reservoir is relatively deep and long, and the spill-over duration is very short, thus almost all the sediment reaching the reservoir is deposited within the reservoir.

4.5 Roma Valley area 5

This is the Roma Valley catchment upstream of the bridge at Lehoatateng on the Roma-Maseru road. The catchment has an area of 57 km² and includes areas 1 to 4 as subcatchments. The descriptions of geology, landforms, climate, vegetation, soils and landuse in Chapter 2 apply to this catchment. In this section only the results of field studies carried out in the study period 1973—1977 are dealt with. Two types of studies were carried out in this catchment for the area not included in catchments 1—4 above, namely: catchment surveys and water and sediment discharge measurements. The results of these studies are described in the following subsections.

4.5.1 Catchment Surveys

In this catchment, the surveys are limited to two field methods:

1. Air-photo interpretation of some gullies between 1951 and later.

Table 4.7. *Notes on Gully Erosion, Landsliding within Roma Valley Area 5: Lower Reaches.*

Gully/Location No.	Type of Damage and Date of observation		Size of the observed damage			Vol-ume
			Length	Width	Depth	
			m	m	m	m³
G1	[1]Scarp retreat	1974	11.63	3.13	1.92	6.1
		1975	7.18	2.72	0.85	2.3
		1977	5.13	2.15	1.53	3.3
G1a	Wall collapse	1974	2.17	3.20	11.61	80.6
G1b	,, ,,	1976	3.51	4.91	10.73	184.9
G2a	,, ,,	1974	5.74	4.12	12.13	286.9
G2b	,, ,,	1975	3.19	1.97	5.09	32.0
G2C	,, ,,	1977	6.81	3.96	7.55	203.6
G3a	,, ,,	1975	9.11	6.31	13.01	747.9
G3b	,, ,,	1977	2.27	1.36	8.19	25.3
L1	Slide scar	1976	21	13	0.94	256.6
L2	,, ,,	1977	Not measured			
L3	,, ,,	1977	,, ,,			

[1] For the scarp retreat the length represents the distance from the bench-mark to the gully-head. The width and depth are measured 1 m downstream of the gully-head and at the gully-head. The two figures represent the mean of the two measurements. The total retreat of the secondary scarp is thus 6.5 m between 1974 and 1977.

2. Annual inventory of erosion damage in the form of bank-erosion, gully-head and gully-side extension.

The field methods yielded very meagre information. Most of the gullies in this area showed very slow or no growth at all during the period studied. Several of the gullies have been dammed since 1968 and this has made it difficult to see any headward extension. Others have reached bedrock scarps and have no headward extension in that direction. However, three main gullies were chosen in 1974 for annual recording of erosion activity. The main characteristics of these three gullies are: extensive width with flat bottoms and almost vertical walls; secondary fill in one of them and secondary incision of the gully fill in the other. Beginning in 1975, these gullies were annually inspected and measurements made at areas of activity in the form of slumps, headward advance of the secondary scarps and side gully formation (Table 4.7).

Another feature noted, though difficult to quantify, was the extent of surface erosion in the area between the photography dates. The three main pediments/spurs in the catchment show progressive increase in pale grey tones from 1951 to 1971. This was interpreted as the result of sheet erosion, mostly on cultivated lands and grazing areas at the lower parts of the spurs. Several of the small rills on the fields seem to have increased and become more permanent, producing a system of depressions extending from the downslope side of buffer strips/contour furrows.

Human activity has increased extensively in the area since the 1950s. The village above the University area has become larger, the University area has been extended and occupies an area double what it was in the early 1950s. The building activity connected with these establishments should have a great impact on the sediment import into the streams. If the construction and tarring of the road to Roma is included, the erosion damage should have been greatly increased.

After the first rains following the author's arrival in Lesotho in 1976, fresh slides were noted on the northern slopes of the catchment. After that date, observations were made on the slopes to record sliding activity. Three fresh slides were observed and one of them, of medium size, was surveyed for size, type of material and volume of the scar left by the removed material (Table 4.7). It was further found that the slopes had experienced several minor slides recently, but most of them had been recolonized by vegetation. All in all, 13 locations with shallow sliding along the bedrock were observed on the slopes. Dates for the occurrence of these are not known but they could hardly have occurred earlier than 1974, as most showed only second generation weeds in the scars.

The main channel is deeply incised into the flood-plain deposits. In the last two kilometres of the reach there are several accumulation terraces. Within this reach sandbanks have formed and at some points these have been eroded and the channel thalweg shifted. On two occasions (1974 and 1977) the whole reach was walked and observations made of lateral erosion and deposition. There was not much difference in the general position of the sandbanks but all the slumped blocks observed in 1974 were washed off in 1977.

4.5.2 Water and Sediment Discharge

In order to make up for the lack of discharge data within the catchment and at the same time provide data for the estimation of gross erosion over the catchment, a gauging station was installed at the new bridge on the Roma-Maseru road south of St. Michael's on the Lehobosi river at Lehoatateng. The staff gauge was first installed in January 1975. The section was surveyed and was found to be lying mainly on bedrock and semi-consolidated materials. The staff gauge was damaged several times during 1975 and 1976 and no complete record could be provided for the period. Spot samples of water were taken during the rainy season 1975/76 (Table 4.8) for the analysis of sediment concentration during the high flows. Very limited data were collected, however, due to lack of personnel during the period. In October 1976 a new staff gauge was installed and an intensive programme of water sampling

Table 4.8. *Sediment Concentration in some Water Samples Collected from Area 5 River Station During some High Flows in 1975 and 1976.*

Date	Sediment concentration mg/l	Rainfall at Roma mm/day
1975-01-24	1,000	22
27	696	25
02-13	783	26
16	391	36
03-01	584	48
1976-01-09	451	4
28	613	27

started. The water stage was read daily and water samples collected three times a week and on as many rainy days as possible. The station was rated five times during the rainy season 1976/77 and the relationship between water stage and water discharge was found to be represented by the expression

$$Q = 7.8 \times 10^3 \, H^{2.6} \qquad (4.3)$$

where Q is the water discharge in l/s, and H the water stage in metres. Discharges corresponding to all observed water stages were calculated using this relationship, and then reduced to m^3/day. The discharges were then ranked and a frequency distribution of the observed discharges obtained (Fig. 4.30).

In order to illustrate discharge variations with time, the computed discharges

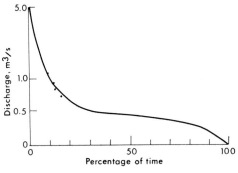

Fig. 4.30. Frequency distribution of water discharge at Lehoatateng river station during the rainy season 1976/1977.

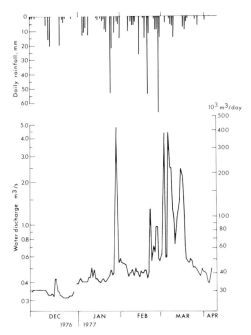

Fig. 4.31. Daily rainfall at Roma, and daily water discharge at Lehoatateng river station, Dec. 1976—April 1977.

were plotted on a time scale with the first day of observation as the zero point (Fig. 4.31). The highest discharge values are concentrated in the period January to March (cf. Figs. 2.4 & 2.5). The two months had 45% of the annual rainfall for the water year 1976/77 (October 1976—September 1977).

A total of 49 water samples were collected during the period, and the sediment concentration determined. Using these data and discharge values for the stages at the time of water sampling, a sediment rating curve was computed (Fig. 4.32). When all the data were included irrespective of discharge range, a relationship between sediment load and discharge was found to be represented by

$$L_t = 13.0 \ Q^{6.9} \qquad (4.4)$$

where L_t is the sediment load in kg s^{-1} and Q is the water discharge in m^3 s^{-1}. The extremely high value for the exponent of Q

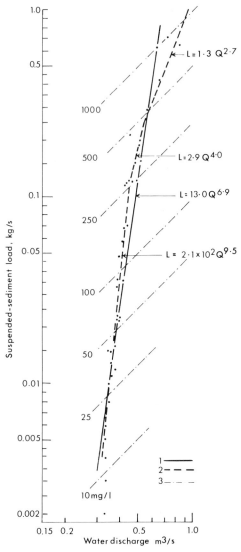

Fig. 4.32. Sediment rating curve, Lehoatateng river station. 1. Calculated relationship between water discharge and sediment load, 2. Curves fitted by eye to the scatter points, 3. Sediment concentration in mg/l.

in this expression is caused by the large variations in sediment concentration in the water discharges below 0.5 m^3 s^{-1}. The variation in sediment concentration in this range seems to be related to other factors besides the volume of water discharge. The scatter of points show three ranges to which three lines have been fitted by eye.

These lines are represented by the equations

$$L = 2.1 \times 10^2 \, Q^{9.5} \quad \text{for } Q < 0.45 \qquad (4.5a)$$
$$L = 2.9 \times 10^2 \, Q^{4.0} \quad \text{,, } 0.45 < Q < 0.54 \qquad (4.5b)$$
$$L = 1.3 \times 10^2 \, Q^{2.7} \quad \text{,, } Q > 0.54 \qquad (4.5c)$$

These equations were used to compute the sediment load for the rest of the water discharges. These were then summed up to give total sediment discharge during the period November 1976—April 1977 (133 days), assuming that the observed discharges represent daily mean discharges. This assumption has the weakness that most of the high flows during the period are possibly not included in the observations as they often occur at night or are so short-lived that they were missed. Furthermore, the majority of low flows were mostly poor in sediment content. The computations give a sediment load of 28400×10^3 kg for the period, or a mean value of 200 tonnes/day. The sediment yield, if these values are assumed representative, would be 1400 t km^{-2} y^{-1}. However, the following should be noted in connection with the representativity of the obtained estimates. The months of June, July and August normally have such low flows that the figure would be lower if this is taken into account. Secondly, the majority of high flows are not included in the point observations used for obtaining these estimates of sediment yield. The two facts should in the long run balance out but as the record is very short in the present context they may not cancel out and therefore are a potential source of error in the values obtained here.

The average bulk density of the soils within the Roma Valley has been found to be 1.14 t m^{-3} (Carroll et al. 1976). This gives a denudation rate of 1.2 mm per year over the whole catchment.

Using a 6-year period, Jacobi (1977), obtained a mean sediment yield of 1979 t km^{-2} y^{-1} for the Little Caledon catchment, corresponding to a denudation rate of 1.4 mm per year. The lower figure in the case of the Roma Valley catchment is probably due to less area under cultivation, and the size of the catchment in the sandstone terrain. In both Jacobi's and Binnie and Partners' reports the basalt areas are found to contribute less sediment than the sandstone catchments.

4.6 Maliele catchment

4.6.1 Introduction

Maliele catchment, with an area of 13.5 km², is located 29° 25′ S and 27° 40′ E in central Lesotho. In Chapter 2 this area is described as part of Roma Valley (Fig. 2.1). The catchment has a maximum length of 7 km, and has three main channels, the largest of which flows in a westerly direction and the two smaller ones in a southerly direction until they meet the main channel some hundred metres upstream of the Mahlabatheng bridge on the Maseru-Roma road NW of St. Michael's Mission (Fig. 4.33). The relief of the area is from 1570 m above m.s.l. at the outlet of the catchment to 1945 m at the highest point in the headwaters (Figs. 4.34 and 4.35). The northern catchment boundary is made up of a series of hills in a NW-SE direction. The southern boundary consists of Cave Sandstone residual hills, degraded spurs on shales and mudstones and sandstone pediment slopes below the Cave Sandstone escarpment.

The surficial deposits in the catchment are mainly alluvium along the main channel and the depressions on the pediment slopes, colluvial materials on the scree slopes in the form of lobes and accumulation in the bedrock depressions. These deposits are heavily encised by a system of gullies (dongas), most of which have either reached the Cave Sandstone escarpment or a local bedrock scarp and thus no longer have any possibility for further headward extension.

The catchment can be divided into the

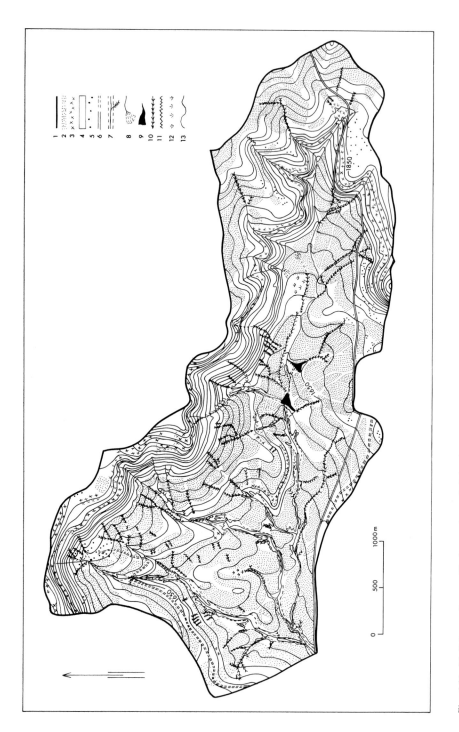

Fig. 4.33. Maliele catchment: Hydrography, relief, land use and gully erosion. **1**. Catchment boundary, **2**. Cropland, **3**. Rockland, **4**. Pasture, **5**. Homesteads, **6**. Roads and tracks, **7**. Hydrography, side banks of entrenched reach (dashed-lines) and side gullies, **8**. Alluvial/colluvial fans, **9**. Reservoir, **10**. Gullies, **11**. Rills, **12**. Swamps, **13**. Contour lines.
Source: Thaba Bosiu Rural Development, 1974: Ortho-photo maps (1:10,000, contoured) & 1971 aerial photography.

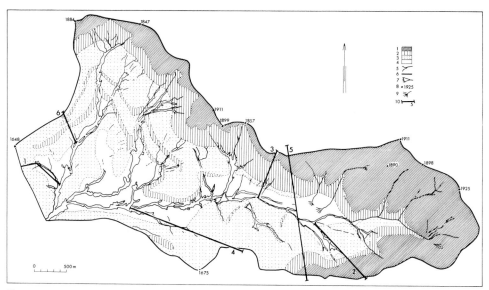

Fig. 4.34. Maliele catchment: Landform zones, gully erosion, and Location of slope profiles. **1**. Basalt Foothills, and Cave sandstone plateau, **2**. Escarpment and scree slopes, **3**. Lowland pediments and spur slopes, **4**. River Valley Flats, **5**. Drainage channels and entrenched zones, **6**. Catchment boundary, **7**. Reservoir, **8**. Spot heights in m, **9**. Alluvial fans, **10**. Location and No. of slope profiles (see also Figs. 4.35 & 4.37a).
Source: Thaba Bosiu Rural Development, 1974: Ortho-photo maps (1:10,000, contoured) & 1971 aerial photography.

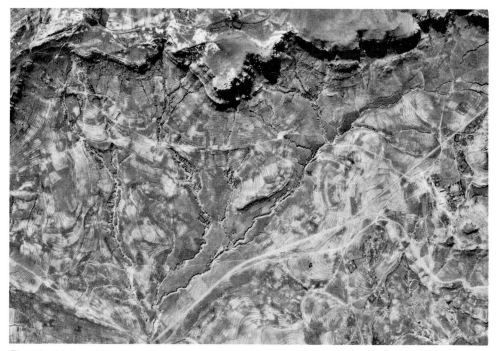

Fig. 4.35. Maliele catchment. An aerial view over the catchment, (May 22, 1961). The cultivation pattern, gully pattern, sandstone rocklands and homesteads are well displayed. The light tones on the cultivated lands are an indication of severe removal of the organic top soil.

93

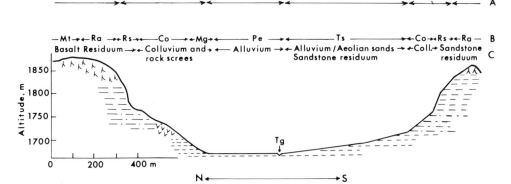

Fig. 4.36. The relationship between landform zones (A), Soil Series (B), topography and geology in a cross-valley profile within Maliele catchment (profile 5 on Fig. 4.35) A. Landform zones (see Fig. 5.34 for explanation), B. Soil Series symbols (see Table 4.12 for explanation), C. Surficial deposits. Tg, indicates location of the channel.
Source: Thaba Bosiu Rural Development, 1974: Ortho-photo maps (1:10,000, contoured) & 1971 aerial photography.

following landforms: basalt hills and dolerite ridges, Cave Sandstone plateau, Cave Sandstone escarpment, scree slopes, pediment and spur slopes and river valley flats or alluvial plain. These are described in section 4.6.2, below where they are used for the summary of erosion features in the catchment, and are grouped into four landform zones. Fig. 4.36 gives a schematic presentation of the landforms across the catchment in a N-S direction.

The soils in the area form topo-sequences of thin light-coloured soils on the hills of basalt and sandstone through colluvium to dark clay loams along the drainage channels (duplex soils). The spurs and pediment slopes are covered with reddish sandy loams (Figs. 4.3 and 4.36). The protected, wet south-facing slopes support some thickets of *Leucosidea sericea*, *Myrsine africana* and *Rhamnus priorides* among other indigenous shrubs.

Large parts of the catchment are under cultivation during the wet season, but are open to grazing of the crop stub during the dry season. The summer grazing is limited to hills and colluvial slopes, along dongas and between the fields. Homesteads are located on the edges of landform zone 1

just above the escarpment and at the foot of colluvial slopes in the lowlands. Some villages occupy the slopes of the residual hills in the lowlands. Land management practices for the catchment are the same as those of the rest of the Roma Valley areas (cf. 4.1—4.5 above). Two reservoirs exist in the catchment, constructed in 1968/69.

4.6.2 Reservoir Surveys

Two reservoirs exist within the main Maliele catchment (Reservoirs 5 and 6). They have catchment areas of 0.60 km² and 6.30 km² with relative reliefs of 210 m and 325 m respectively. They were constructed in 1968/69, but no data exist concerning their original volumes after completion. Both reservoirs have earthen embankments and overflow spillways. In addition, reservoir 5 has a valve gate at the lower centre of the embankment. Reservoir 6 had originally a concrete crest overflow spillway, but this has been separated from the earthen embankment by a break at the contact zone, resulting in lowering of the full supply level of the reservoir by 50 cm.

The reservoirs were sounded twice

Fig. 4.37a. Ground view of landform zones 1 (Horizon), 2, 4 and 3 (Foreground). Note also the location of discontinuous gullies at the boundary zone between landform zones 2 and 4, and the stoniness, terraces, slide scars and rock boulders on the scree slopes. (Photo: Q. K. Chakela, Nov. 1974.)

during the period of study, in 1974 and 1977. The bottom sounding data were used to produce bottom maps of the reservoirs (Figs. 4.40—41).

Volumes of the reservoirs at the two surveying occasions were computed and reservoir capacity curves drawn using the volume data (Fig. 4.42, Table 4.9). In addition, the differences in depth at the sounded points between 1974 and 1977 were used to produce depth-difference maps of each reservoir (Figs. 4.40c—41).

Fig. 4.37b. Gully formed in a topographic depression within landform zone 3, Maliele catchment. Note the side pipe opening into the gully to the left. The pipe continues to beyond the ant-hill. (Photo: Q. K. Chakela, Feb. 1975.)

Table 4.9. *Reservoir Volumes, Sediment Accumulation and Rate of Sedimentation in Maliele Reservoirs, 1974 to 1977.*

Reservoir No.	Reservoir volumes		Sediment Accumulation			Rate of Sedimentation per unit area of reservoir
			Volume	Weight		
	1974 m³	1977 m³	m³	wet tonnes	dry tonnes	cm/year
5	22,276	21,266	1,009	1,433	324	5
6	7,313	5,979	1,334	1,894	429	9

Core samples were collected from some longitudinal lines through the reservoirs for the study of sediment texture and structure by visual inspection in the field. Some of the core samples collected from reservoir 6 in 1977 were analyzed for wet bulk density, particle density, grain size composition and water content (Fig. 4.44a). Some sediment samples were also collected from the sandbanks of reservoir 6 for analysis of grain size composition (Fig. 4.44b).

Manual staff gauges were installed at the reservoirs in 1974 but were vandalized several times during the rainy seasons of 1974/75, and therefore no reliable record could be obtained. More securely fixed and hidden gauges were installed at the beginning of the 1976/77 rainy season and read daily in connection with water sampling and stage readings at the Mahlabatheng river station. These data were used to produce water level records of the reservoirs for the 1976/77 field period (Fig. 4.43).

Water samples were collected at the outlet of reservoir 6 during the 1976/77 rainy season and sediment concentrations were determined. These values are compared to the rainfall amounts recorded at Roma and Khokhotsaneng rainfall stations.

In short, the reservoir surveys within Maliele catchment consisted of bottom sounding of the reservoirs, determination of reservoir volumes, bottom core sampling for the study of reservoir deposits, sampling sandbanks in the vicinity of the reservoirs for the study of the morphology and texture of the deposits above the reservoir's full supply level, stage recording of the reservoirs and, where possible, sampling water leaving the reservoir for sediment concentration analysis.

4.6.3 Catchment Surveys

Catchment surveys within Maliele catchment included the following:

1. Air-photo interpretation of the aerial photographs taken in 1951, 1961 and 1971. The main objective was to study the intensity and distribution of the different erosion processes in the area. Gully erosion is the visually most dominant feature in the catchment, therefore some gullies were selected for comparison of their sizes on the different photographs and to get a measure of their growth rate (Table 4.5).

2. Ground measurement of gully growth by installation of pegs above some gully-heads and selection of some gully cross-sections for repeated measurement. The data obtained in this way are presented in Table 4.5 together with data obtained from aerial photo comparison.

3. Transects and slope profile surveys were carried out in some selected areas. Most of these surveys were made along side gullies to the main stream of the catchment. During the surveys, notes were taken concerning slope gradients every 20 m or 50 m interval, vegetation type and cover for each 20 m interval, soils, bedrock, land use and gully depth for each interval. Slope profile surveys along gullied reaches included cross-profiles,

surface and bottom of the gully longitudinal profiles (Figs. 4.38 & 4.39).

4. At the end of each field period, and occasionally after rainstorms, observations were made along the major gullies concerning lateral extension of the side walls of the gullies, changes in the head-scarps of the side gullies and areas of deposition and erosion along the bottom of the gullies. Sediment samples were taken from some of the sandbanks found in certain reaches along the main stream between

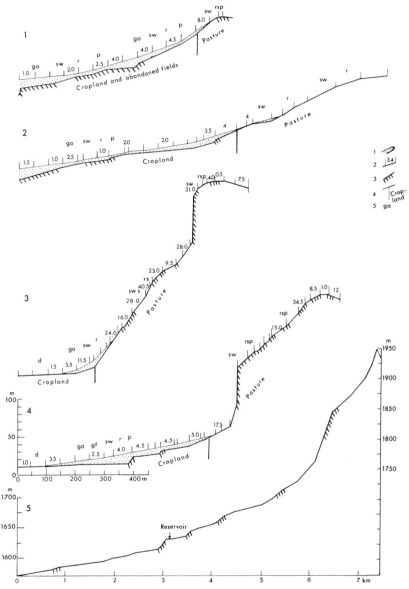

Fig. 4.38. Slope profiles, and longitudinal profiles of some gullies and the main stream channel of Maliele catchment. Diagrams 1, 2, 3, 4 correspond to 4, 3, 2 marked on Fig. 4.35 and measured in the field. Diagram 5 is the longitudinal profile of the main stream channel. **1.** Gullied reach, **2.** Measured slope section and gradient in degrees, **3.** Bedrock outcrop, **4.** Land use boundary, **5.** Observed erosion processes along the profile (see Fig. 4.16 for explanation of the symbols).

97

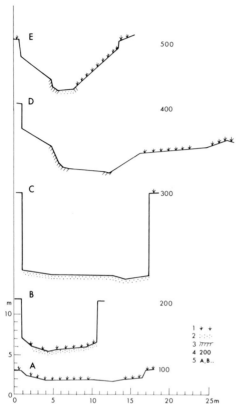

Fig. 4.39a. Transverse profiles across the gullied reach of profile 4 on Fig. 4.38. **1.** Grassed sections of the profile, **2.** Recent, loose deposits (see Fig. 4.39b for texture), **3.** Bedrock outcrops, **4.** Distance of the profile location from the gully mouth, **5.** Profile identification letters.

neglects the morphological elements covered by each slope gradient classes.

Landform zone 1: Basalt hills and sandstone plateau

This landform zone encompasses the Foothills and Escarpment land systems of Bawden and Carroll. It represents a structural level zone above the Cave Sandstone escarpment and is underlain by sandstones and basalt lavas and criss-crossed by several ridges of dolerite. The relief of the zone is 145 m (1800—1945 m). It forms the highest part of the area. Gullying is very rare and limited to drainage depressions and along footpaths and cattle routes. An aerial photo interpretation of the area reveals three types of erosion: bare surfaces around villages without any vegetation cover; exposed bedrock surfaces along the edges of the escarpment and sometimes extending into basalt terrain; rills on the fields and several parallel rills along footpaths between villages and fields or between villages. On aerial photographs gullies are difficult to distinguish from stream channels in this zone and it was only through ground checking that the gullies measured on the aerial photographs could be distinguished from the streams.

Ground checking in the field confirmed the erosion features seen on air-photographs. The following features could be observed in addition to those described above: two types of overgrazing occur in the zone, complete uniform overgrazing of the grass to heights of some centimetres leaving the grass as a well-clipped lawn and without any bare ground patches in between; secondly, overgrazing which has led to isolated bushes separated by bare ground and total absence of palatable grasses. The invasion of the area by Karroid bushes along the hills and slopes on abandoned fields, ephemeral formation of rills on the fields, and the stripping of the regolith along the edges of the sandstone

reservoir 6 and the bridge at the outlet to the catchment, and subjected to size analysis (Fig. 4.39b).

In order to simplify the presentation of the results of the catchment surveys, the catchment was divided into four landform zones based on form, geology and soils as revealed on aerial photographs and through field checking (Fig. 4.34). The three land systems of Bawden and Carroll (1968) represented as covering the catchment were found to be too broad, particularly for the description of the Central Lowlands Land System. The soil map of the area by P. H. Carroll et al. (1976) is very general in its slope groupings and

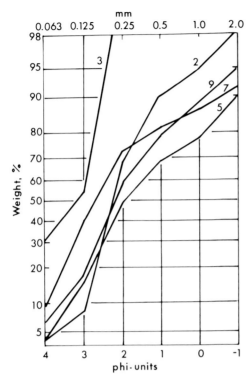

Fig. 4.39b. Grain size composition of gully-fill deposits from gully 2, Maliele catchment. Samples 2 and 5 are collected along the centre of the active channel, at 100 m, 300 m and 370 m (below a bedrock scarp) from the gully mouth. Samples 7 and 9 were collected on either side of the active channel at 370 m.

bedrock outcrops forming a retreating scarp 10—30 cm thick, are some of the features observed in this zone.

The vegetation in this zone is dominated by grasses and small shrub thickets along the channels and on wet south-facing slopes. The soils are mostly lithosols on bedrock hills and sandy to fine sandy loams grading to clay loams along the footslopes and depressions. The soils in the whole catchment have been mapped into soil series and are given in Table 2.3.

Landform zone 2: Escarpment and Colluvial/scree slopes

The zone consists of the Cave Sandstone escarpment cliff and steep slopes, scree

slopes below the escarpment extending either to the main stream alluvial plain or ending above pediments or spur slopes (Fig. 4.37). The top of the Cave Sandstone is generally at 1800 m and forms a continuous cliff line broken here and there by drainage channels and hollows along dolerite dykes. The zone thus comprises of two distinct units: a scree slope with gradients between 15° and 35°, straight in form but occasionally concave upwards and strewn with large boulders on colluvial lobes and a steep escarpment forming vertical cliffs with caves in several places. There are several landslide scars on the scree slopes but only two showed activity during the present field studies. The cliff shows no recent activity, but the boulders scattered on the colluvial lobes testify to past active rockfalls from the cliff. There is another minor escarpment at about the 1670—80 m level running parallel to the main escarpment. Here the boulders and stones are not distributed linearly downslope but are heaped in clumps. In discussing these rock heaps in South African literature some of them are attributed to lightning strikes (King, 1951). My field studies in this catchment and observations in the Roma Valley and Khomo-khoana catchments indicate, however, that the hard sandstone cap is underlain by easily weathered shales and mudstones which are removed by water. The top rock collapses and shatters into pieces of different sizes, in some cases as rock slabs a square metre or more. This is a more probable explanation of the rock heaps. Between the lower escarpment and the main scree slope is a gently sloping surface of the third landform zone described below.

The soils in this zone are mostly unsorted colluvial materials with thick layers of sandy to fine sandy loams separated by thinner dark to black layers. The thickest colluvium is found in the drainage depressions which seem to be formed on the underlying bedrock at the head of most streams in the lowlands.

Gullying is extensive in these drainage channel depressions and most of the gullies trenching the catchment start in this landform unit where most of them have reached the bedrock escarpment. The gullies have trenched the thick colluvial deposits to the bedrock and are now eating into the softer mudstones of the Red Beds, producing stepped gully bottom profiles. All the gullies in this unit are displayed well on the aerial photographs.

There is very little cultivation in the unit due to slope steepness and thin stony soils on the interfluves and gullying in the depressions. The rest of the area is grazing lands and villages. The south-facing slopes support thick local shrub thickets and some planted poplar trees near the village of Khokhotsaneng.

Although gullying is the most visible form of erosion in the catchment and in this landform unit, surface wash by overland flow on the interfluves seems to be the dominant form of erosion due to intensive overgrazing. This has led to increased stoniness and abandoning of many fields which now stand bare with a hard crustal surface which does not support any vegetation.

Interpretation of the 1951, 1961 and 1971 aerial photographs reveals small changes in gully form, size or pattern. However, field studies during the rainy seasons of 1974/75, 1975/76 and 1976/77 showed that gully collapse and lateral extension in this zone do exist and initiation of side gullies along contour furrows is taking place (Table 4.5).

Landform zone 3: Pediments and spurs

A gently sloping plane extends from the foot of the scree slopes at the end of landform zone 2 to the flat area along the main drainage channel. It forms part of Bawden and Carroll's Central Lowlands land system. It is composed of rounded spurs separated by drainage depressions with thick colluvial/alluvial deposits which are trenched by gullies, most of which show multiple trenching. The slope gradient towards the main channel varies from 5—10°.

Vegetation is mainly grasses and crops on cultivated lands, some of the fields having been abandoned due to surface wash. All the main gullies in this unit continue into the upper landform unit. There is extensive cultivation in the area and pastures are limited to spur-end slopes, dolerite hills and along gullies. Wash erosion is more pronounced in this zone than in all the other zones in the area. It shows itself in the form of "sand ripples" on interfluve slopes after rain storms, earthen pedestals and bush pillars, and rills on the fields immediately after ploughing. The majority of the main gullies in this zone show piping, whereby pipes either run parallel to the gullies for several tens of metres or empty into the gullies either at the bedrock/regolith interface or above a clay pan or a concretional layer in the soil profile. Reservoir 5 is situated in this landform zone and the changes in its bottom profiles seem to be related to piping of the form shown in Fig. 4.37b, the collapse of which caused apparent deepening of the reservoir along the sides.

Landform zone 4: River Valley Flats

This is a flat zone with slope gradients varying from 0—2%. It occupies the alluvial plain of the catchment and its main tributary channels. The soils are mostly alluvial soils and duplex soils where colluvium or alluvium overlies the older sediments. These soils are highly sensitive to gullying. The boundary between this zone and zone 3 is a very diffuse break in slope and most of the tributary parts of the zone extend along the depressions into the pediment and spur zone.

The area is heavily cultivated and pastures are limited to gully bottoms, gully sides, and between fields and footpaths. Three swampy areas exist in this zone. One is above reservoir 6 north of Boinyatso village. Up to the drought of 1970 this area supported a rich reed meadow but now it is almost drained off. Two minor swamps or areas of high ground water level exist on the northern side of the main stream and they are both being drained off by a system of gullies with piping.

There are very few sandbanks in the main channel except those related to the damming effect of reservoir 6. The minor sandbanks on the lower reaches of the main channel were sampled for sediment texture analysis. They were in most cases gravelly and stony with little fines. This seems to be related to the competence of the stream for fines, leaving the gravels to be transported as bedload during floods. Fig. 4.48 summarizes the grain size distribution of these sandbank deposits.

4.6.4 Water and Sediment Discharge

In order to obtain an estimate of the amount of sediment leaving the catchment, a water sampling and stage recording station was established at Mahlabatheng bridge on the Roma-Maseru road. The station was selected in 1973 and point measurements of sediment concentration were made after some rainstorms in the rainy seasons, 1974/75 and 1975/76 (Table 4.10).

Water level gauges installed during this period were vandalized several times, so the record of water stage is inaccurate for the estimation of water discharge and sediment transport and is not comparable with that of the rainy season 1976/77, because the gauges never remained long enough for them to be surveyed into the permanent points and the temporary points to which they were tied were also vandalized.

In November 1976, a manual stage gauge was installed at the bridge and tied to both bridge pillars, bottom (maximum depth) and a local bench mark. The cross-

Table 4.10. *Spot Samples of Sediment Concentration at Mahlabatheng River Station, Maliele Catchment*

Date	Sediment concentration mg/l	Remarks
1975-01-24	1,067	Heavy afternoon storm, water muddy. Crab sample.
1975-01-27	3,832	High water after a night rain. Sample taken during a local storm.
1975-12-17	4,270	Heavy storm only in the Maliele catchment and not recorded at Roma.
1976-04-14	283	Sunny day, low water.
1976-04-23	297	Sunny day, low water.

section was surveyed with a level tube. The section partly is underlain by concrete and bedrock and has no depositional points upstream of the measured line. The section was rated five times and at each rating the bottom was checked for possible deposition. The relationship between discharge and stage was found to be represented by equation

$$Q = 4.4 \, H^{1.6} \qquad (4.6)$$

where Q is water discharge in $m^3 \, s^{-1}$ and H water stage in m. The above relationship was used to calculate discharge for all stages recorded, including readings taken in connection with water sampling for sediment concentration analysis. These discharges were then grouped and the mean discharge calculated for all the observations. The mean discharge for the rainy season was found to be 129 l/s.

It should be noted, however, that these figures represent point measurements in time and not the continuous stage-discharge relation, due to the flashing nature of the flows after rainstorms. The only peak discharge that appears to be included in these observations is the one which occurred on 5 March 1977. This was, however, less than the stage during the night of 2—3 March, following a rainfall which was recorded as 67 mm at Roma and 70 mm at Khokhotsaneng rainfall stations respectively. These two rainfall records are the highest recorded since the beginning of these studies. The flood of 5 March 1977 passed within two hours, with a drop from 51 to 18 cm in that period. The frequency of observed values of discharge (Fig. 4.46) represents therefore a partial time-frequency relationship as most of the high discharges are probably not included in the observations of stage.

Forty-six water samples were collected below a minor rapid upstream of the bridge at low water and downstream of the old bridge at higher flows, using a time intergrating, hand-operated sampler (Fig. 3.2). The samples were treated as described in Chapter 3. Sediment concentration and load were determined.

A relationship was determined between water discharge and sediment load for the 46 samples and is represented by the equation

$$L_a = 2.0 \; Q^{2.2} \qquad (4.7)$$

where L_a is suspended-sediment load in g/s and Q is water discharge in l/s. This relation was used to calculate sediment load for the rest of water discharge values for which no water samples were taken.

In order to study the effect of rainfall on sediment transport indicated in spot samples taken during the 1974/75 and 1975/76 rainy seasons, a sediment rating curve was calculated using only those data corresponding to rainy days. The rainfall record used is that of the Roma rainfall station and it should be borne in mind that these rainfall data do not correspond exactly with the occurrence of rain within the Maliele catchment and that it is for a whole day. For example, the record for the 15th represents the rain that might have occurred between 8 a.m. on the 15th and 8 a.m. on the 16th or 2 p.m. and 2 p.m. on

the respective dates depending on when the rain gauge was emptied. The water samples were taken between 12 noon and 3 p.m. In addition, the rainfall which produces high flows in Maliele may not be recorded at Roma due to the fact that it may be very local. The equation obtained using these data was found to be

$$L_r = 2.4 \; Q^{2.4} \qquad (4.8)$$

Using the equation established above, sediment transport was calculated for the rest of the rainy days in the period November 1976—April 1977.

Furthermore, a regression line was determined using only those samples with discharges greater than 50 l/s. It was found to be

$$L_{50} = 2.7 \; Q^{2.5} \qquad (4.9)$$

These three curves are represented in Fig. 4.47, together with the scatter of observation data. The three curves lie very close to each other and are within the standard deviation of the observed values.

From these analyses it was concluded that the regression line determined using all the data was representative for the data during the period of observation. The mean daily sediment transport was calculated for the whole period and for the rainy days of the study period. These values were found to be 9.70 and 15.79 t/day respectively. If the former figure is used for estimation of annual sediment yield, a value of 270 t km^{-2} y^{-1} is obtained. The above calculations lead to the conclusion that during the 1976/77 rainy season, 65% of the sediment was transported during 40% of the time and by 44% of the total water discharge during the period.

4.6.5 Results and Discussion

Land use within the catchment is predominantly cultivation followed by pastures. Over 60% of the lowlands are intensively cultivated. This has resulted in land deterioration and bad land use practices such as cultivation on very steep slopes encroaching on pastures and therefore increasing the livestock pressure on the grazing lands, which in turn has led to increased erosion and overgrazing followed by increased sheet wash on pastures.

Landforms in the catchment are hills and hillcrests of dolerite and basalt, sandstone plateau, sandstone escarpment, scree slopes spurs and pediment slopes and a narrow alluvial plain in the centre of the catchment. In this report these landforms have been grouped into four zones.

Dominant erosion features within the catchment are *gullies* (dongas) which are mainly of two types: valley-bottom gullies and valley-head continuous gullies which are mainly the continuation of the entrenched drainage channels, with secondary scarps formed by bedrock terraces; valley-side discontinuous gullies, several of which terminate at the alluvial plane before reaching the main streams. Some of the former tributaries of the main stream have been cut off at the confluence by alluvial fans from the gullies and transforming these into discontinuous side gullies. The largest of the discontinuous gullies are erosional in the upstream reaches and depositional in the lower more gently sloping sections and have maximum depths in the mid-gully reaches. Thus a large proportion of the sediments channelled through these side gullies never reach the main stream. The rate of gully growth for the active gullies, was found to be erratic both in space and time. The annual rates of the longitudinal extension were found to be in the range 0—4 m (Table 4.5) and lateral growth of a few gullies was observed to be ranging from a few decimeters to 5 m. The slump blocks from lateral erosion rarely leave the erosion sites and are often stabilized in position by vegetation. Several of the gully walls were observed to be colonized by vegetation. Vegetation colonization has also extended to sandbank deposits along the gully bottoms and most of the flat-bottomed reaches have been completely covered by vegetation, which is acting as a trap for the sediments reaching these parts of the gullied reaches. Gully wall slumping seems to be the major factor in initiation of local deposition within the gullies.

Besides the spectacular gully erosion feature, catchment surveys revealed that *surface wash and rill formation* are important erosional processes on the interfluve areas on spurs, pediment slopes and scree slopes, reaching maximum intensity on cultivated fields on gradients higher than 6°, and around villages where they are aggravated by livestock activity. Where surface wash and rill formation has been extreme (on side slopes of lowland spurs and pediments) the top soil (organic layer) has been completely removed leaving either bare ground with pale-coloured soils, crustal surfaces or bare bedrock. In some cases whole fields have been put out of cultivation. How far this is caused only by surface wash and raindrop erosion or the combined effect of bad cultivation practices and raindrop erosion is not clear.

Slide scars, "Sheep tracks", colluvial lobes and boulders on the scree slopes testify to past very active *mass movements* processes within the catchment. However, at present mass movements are limited to rounded minor valley heads on south-facing slopes in the scree slopes/escarpment landform zone.

Subsurface piping is yet another erosion feature encountered within the Maliele catchment. Most of the gully-heads seem to extend upstream through underground tunnels. In some cases the pipes are observed as holes in the side gullies, and where these have fallen in, a new side gully is initiated. In some locations pipes could be observed only as holes in the ground forming a linear alignment tributary to the gullied reach but not opening into the gully. Several of the small depressions with

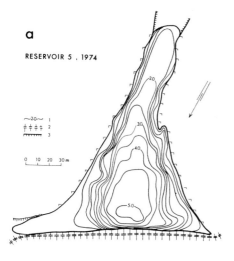

a

RESERVOIR 5 , 1974

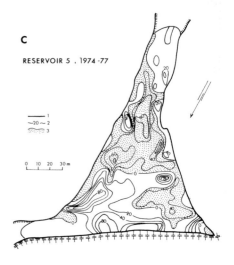

c

RESERVOIR 5 , 1974-77

c) **1**. 1974 full supply level, **2**. Depth difference iso-
line in cm, **3**. Area of negative depth differences
(erosion).

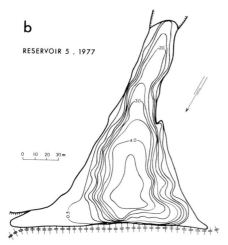

b

RESERVOIR 5 , 1977

Fig. 4.40. Bottom Maps of reservoir 5. a) Nov. 1974,
b) March 1977, c) Depth differences between 1974
and 1977. a) **1**. Contour line in m below full supply
level, **2**. Reservoir embankment, **3**. Gully scarps
above full supply level.

swampy areas in the lowlands were
observed to be undergoing draining
through pipes.

Reservoir sedimentation in the Maliele
catchment is very rapid. Reservoir 6, constructed in 1968/69 had lost over 10 m of
its depth through sedimentation by 1974,
and very little material seems to be retained in the reservoir during the study
period.

Reservoir 5, on the other hand, shows
very little loss of depth compared to the
original gully depth. This may be due to
the fact that the reservoir was deeper than
the gully. Mean annual sediment accumulations in the two reservoirs between
1974 and 1977 were found to be 432 m³
and 572 m³ respectively for reservoirs 5
and 6. The median grain size of these reservoir accumulations and of the deposits
found above full supply level fall in the
range 0.18—0.38 mm and 0.08—0.80 mm
respectively with silt-clay fractions of up to
45% (Fig. 4.44). The mean grain size
(Folk and Ward, 1957) was calculated to
be 1.7 mm and dry bulk density to 0.67 g
cm⁻³, respectively. The sediment shows
varving with alternating layers of fine and
coarse grained materials. The close
proximity of the layers and their number
per decimetre were interpreted to be due
to differences in inflow rates during the
same rainy season rather than to represent
annual layers. The coarse sediments were
brought into the reservoirs during heavy
storm water flows, and the fine material
was deposited during more calm conditions with low inflow rates.

Sedimentation within the catchment does
not occur only in the reservoirs. Three

104

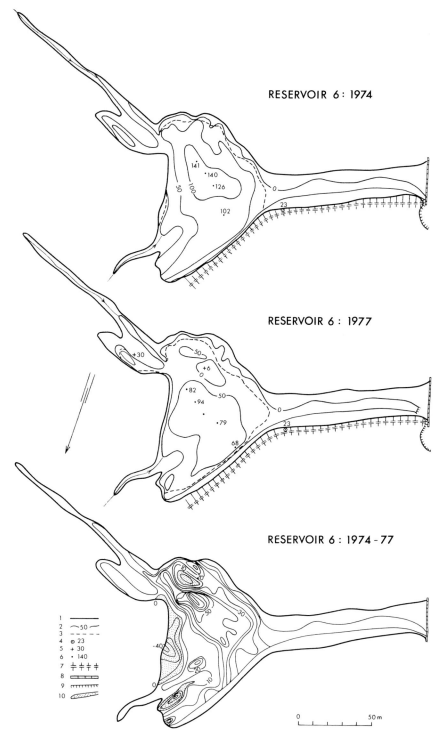

RESERVOIR 6 : 1974

RESERVOIR 6 : 1977

RESERVOIR 6 : 1974 - 77

Fig. 4.41. Bottom maps of reservoir 6. a. Nov. 1974 b. March 1977 c. depth differences between 1974 and 1977.
1. Original full supply level, **2**. Contour line in m below full supply level (for a) or depth difference iso-line (for c.), **3**. Water-vegetation boundary, **4**. Bench-mark cm above full supply level, **5**. Spot heights in cm above full supply level, **6**. Spot depths in cm below full supply level, **7**. Reservoir embankment, **8**. Concrete spillway, **9**. Actively eroding gully scarp, **10**. Zone of local erosion (negative depth differences).

Table 4.11. *Summary of Discharge and Sediment Load Calculations for Mahlabatheng River Station, Maliele Catchment*

		Water Discharge m³	Sediment Load kg
No. of days covered	163	2,665,653	1,582,330
No. of days with rain	65	1,182,657	1,026,290
Rainy days for w.yr. 1976/77	95		

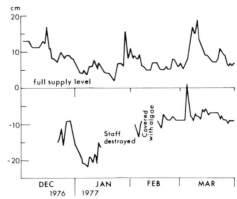

Fig. 4.43. Water level record, Reservoirs 5 and 6, during the rainy season 1976/77.

other depositional zones are: low gradient reaches within the main gullies, at the foot of the scree slopes where the gradients change to lower values, and finally at the mouths of discontinuous gullies and small streams in the form of small alluvial fans.

Reservoir deposition represents, in general, a fraction of the net erosion in the catchment areas of the reservoirs and can be used as a measure of sediment yield from the areas. Within Maliele catchment, sediment accumulation in reservoir 5 can be taken to be a good estimate of the sediment yield from the catchment. The reservoir rarely overflows, and when this happens, the spill-over discharge and load is very insignificant compared to the inflow discharge and load. Therefore, using the figure obtained from the reservoir's volume decrease between 1974 and 1977, a sediment yield estimate of 220 t km^{-2} y^{-1} is obtained for the catchment area of reservoir 5. In the case of reservoir 6, the sediment accumulated in the reservoir

between 1974 and 1977 greatly underestimates the sediment yield from the reservoir catchment area. The size of the reservoir at the start of the surveys had been gravely reduced in relation to the sediment delivered to it by the feeding streams.

The result was severe reduction in trap efficiency, making the sediment accumulated in the reservoir between 1974 and 1977 a very small fraction of the materials delivered to the reservoir. In addition, several depositional zones exist upstream of the reservoir where large quantities of erosion products are deposited. The sediment accumulation corresponds only to 20 t km^{-2} y^{-1} of sediment yield during the study period. In order to obtain a better estimate covering the life of the reservoir to

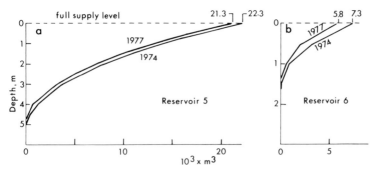

Fig. 4.42. Capacity curves for reservoirs 5 and 6. The figures 21.3, 22.3, 5.8 and 7.3 refer to full supply capacities for the indicated years.

Table 4.12. *Soil Series Names and Classifications. The Symbols are Used in Figs. 4.3 and 4.36.*

Symbol	Classification according to:			
	P. H. Carroll et al. (1976)	Carroll and Bascomb (1967)	Binnie & Partners (1972)	U.S.D.A. Taxonomy (Subgroups)
Be	Berea series	Fersiallitic soils, Berea set	Sandstone soils, Berea series A	Aquic Hapludolls
Fs	Fusi series	Calcimorphic soils, Mokhotlong set.	—	Cumulic Hapludolls
Kh	Khabo series	—	Alluvial soils, Khabo A series.	Typic Argiudolls
Le	Leribe series	Fersiallitic soils, Leribe set	Red soils, Leribe series A	Typic Hapludolls
Ma	Machache series	Eutrophic Brown soils, Machache set	Dolerite soils, Matšaba C series	Typic Argiudolls
Me	Maseru series	Claypan soils, Maseru set	Duplex soils	Typic Albaqualfs
Mg	Maliele series	—	—	Cumulic Hapludolls
Ms	Matšana series	Calcimorphic soils, Mokhotlong set	—	Typic Hapludolls
Mt	Matšaba series	Eutrophic Brown soils, Matšaba set	Dolerite soils, Matšaba "A"	Typic Argiudolls
Nt	Ntsi series	Lithosols on sedimentary rocks	Lithosols	Lithic Udorthents
Pe	Phechela series	Vertisols of topographic depressions, Thulo set	Dolerite soils, Phechela series	Typic Pelluderts
Ra	Ralebese series (Mountain slopes)	Lithosols on rocks rich in Fe-Mg minerals	—	Lithic Hapludolls
Rs	Sandstone rocklands	—	—	—
Rv	Riverwash	—	—	—
Se	Sephula series	Claypan soils	Duplex soils, Sephula series	Aeric Albaqualfs
So	Sofonia series	Juvenile soils on riverine alluvium	Alluvial soils, Caledon B series	Fluvenetic Hapludolls
Ta	Thabana series	Vertisols of Lithomorphic origin, Semonkong set	—	Aquic (Typic) Argiudolls
Th	Theko-Thoteng	—	—	Typic Haplaquents
Ts	Tsiki series	Claypan soils, Maseru set	Duplex soils	Typic Albaqualfs

1974, some cross sections of the reservoir gully reach were measured on air-photographs of 1961 and then adjusted to full supply level of the reservoir before the embankment break. The total sediment estimate was found to be 16370 m³ for the period 1969 to 1974. This corresponds to a mean sediment yield of 350 t km^{-2} y^{-1}. When considering this figure it should be borne in mind that the period 1969—71 was extremely dry and therefore very little material should be expected to have reached the reservoir at that time. Calculation of materials likely to be deposited in the reservoir during the period 1974—1977 (Fig. 4.45) shows that the amount of fine sediment deposited during the spill-over period is very little and varies insignificantly with changes in inflow rates. The calculations are based on the assumption that no re-suspension takes place. However, this is an underestimate due to the shallow depth of the reservoir in the study period. Very little sediment with grain sizes coarser than 0.06 mm leaves the reservoir in suspension. The estimated indices of sedimentation for reservoir 6 indicate that reduction of the inflow rate from 1 m³/s to 50 l/s increases the trap efficiency of the 1974 pool from 30% to 85%.

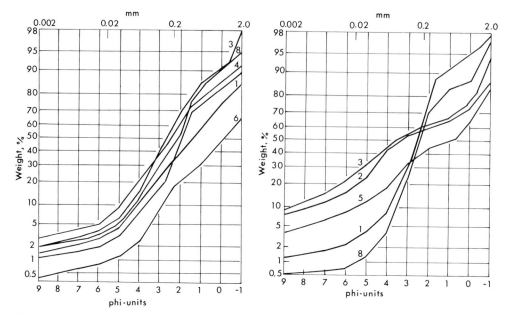

Fig. 4.44a. Grain size composition of sediment samples from the sandbars at the mouth of the main stream channel feeding reservoir 6. Sample numbers indicated relative location of sampling points upstream of the reservoir.

Fig. 4.44b. Grain size composition of core samples from the bottom of reservoir 6. Sample numbers show the location of the sampling points from the reservoir embankment towards the main inflow stream channel.

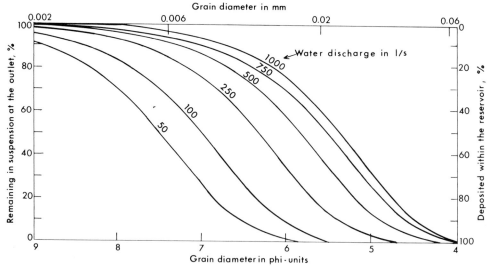

Fig. 4.45. Relative trap efficiency estimates of reservoir 6 based on Sundborg formula. The calculations are based on water temperature of 20°C, sediment particle density of 2.65 g cm^{-3}, 20% (1570 m^3) of the 1974 reservoir volume is taking part in through flow, and that no resuspension took place within the reservoir.

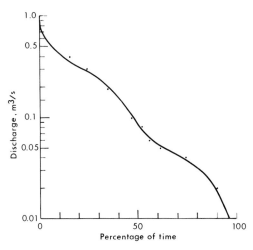

Fig. 4.46. Frequency distribution of water discharges at Mahlabatheng river station during the rainy season 1976/77.

Taking these facts into account, plus the observations on the types of flows during the rainy seasons and sediment concentrations observed at the outlet it is evident that reservoir sedimentation between 1974 and 1977 is a very poor estimate of net erosion in the area and the rough estimate obtained through original volume reconstruction is a better estimate of sediment leaving the catchment.

The reservoir survey results can be summarized as follows: The rates of reservoir siltation are high and loss of capacity rapid. They give sediment yield estimates of 200—400 t km^{-2} y^{-1}. The sediment accumulations in the reservoirs within the catchment do not strictly represent net erosion in the catchment, especially reservoir 6 since 1974. Reservoir 6 will probably be completely filled within two years and the water supply stored in this reservoir will thus have been completely lost to live-stock within a period of less than ten years. Present deposition in the reservoir seems to be mainly encouraged by vegetation colonization around the reservoir.

Water and sediment discharge studies were concentrated to the rainy season 1976/77 and the findings reported apply mainly to

this period. The aim was to estimate two parameters for the whole catchment: First, mean water discharge for the rainy season and then using it to estimate an annual value of runoff from the catchment. Secondly, to estimate the mean sediment load for the rainy season and use it to obtain an estimate for total sediment yield of the catchment. The fact that these data are based on only one rainy season makes the estimates very poorly representative of the annual values. The sediment load figures should be considered in relation to the values obtained for reservoir catchments. The grain size composition of the materials deposited in the reach between the river station and reservoir 6 is generally coarser compared to those deposited in the reservoirs and upstream of the reservoirs (Figs. 4.44 and 4.48).

Mean water discharge for the rainy season was found to be 130 l/s and a total of 2700×10^3 m^3 for the rainy season. This

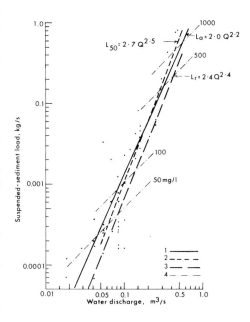

Fig. 4.47. Sediment rating curve, Maliele at Mahlabatheng river station. 1. Regression curve obtained using all measured sediment concentration values, 2. Regression curve for discharges greater than 50 l/s, 3. Regression curve for rainy days only, 4. Sediment concentration in mg/l.

109

gives an annual runoff of about 300 mm from the catchment under the assumption that the mean water discharge is the same for the dry season. The total transport load during the 1976/77 rainy season (163 days) was found to be 1580 tonnes. This gives a sediment yield estimate of 270 t km^{-2} y^{-1} for the water year 1976/77. The relationship of these values to other years and seasons is complex and depends mainly on the intensity and quantities of rainfall during the individual rainy seasons and changes in the land use patterns in the catchment. The fact that the water discharges were estimated from manual readings of staff gauges, and that most of the floods with high sediment loads occurred during the nights and are possibly not included in the computations, would weigh against the fact that in estimating the annual figure the dry months were taken to be contributing the discharges. It is possible that also this extrapolation underestimates the real sediment transport away from the area.

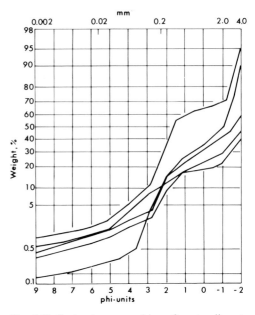

Fig. 4.48. Grain size composition of spot sediment samples collected on the sandbars between reservoir 6 and Mahlabatheng river station.

5. Soil erosion: Khomo-khoana catchment

5.1 Khomo-khoana catchment area 1

5.1.1 Introduction

Khomo-khoana catchment area 1 is a small area of 0.7 km² located within the Khomo-khoana catchment described in Chapter 2 (Fig. 2.15). The investigated area lies on the southern slopes of Muela mountain and is bounded on the south by the Khomo-khoana river (Fig. 5.1). It has

a maximum width of 750 m and maximum length of 1.5 km. The side slopes are washed by several ephemeral streams, few of which have definite channels. The relief of the area is 190 m ranging from 1560 m at the river to 1750 m at the top of the highest peak of Muela mountain.

The landforms in the area include degraded escarpments, scree slopes with large colluvial lobes covered by boulders and pebbles, narrow pediment slopes and an alluvial plain. The river has cut into the

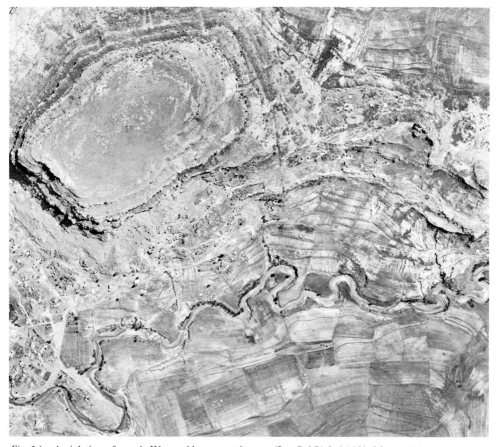

Fig. 5.1a. Aerial view of area 1, Khomo-khoana catchment. (Les Col 71:2, 5:1569, May 1971.)

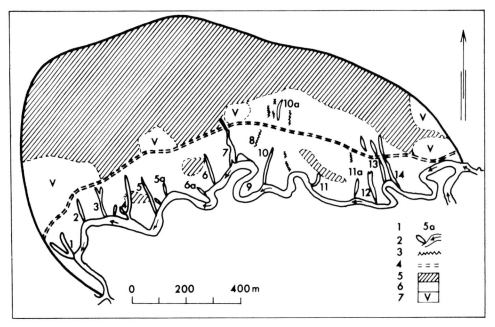

Fig. 5.1b. Sketch map of area 1, Khomo-khoana catchment based on air-photo interpretation of aerial photographs taken in 1952, 1961 and 1971. **1**. Surveyed gullies, **2**. Main stream channel and side gullies, **3**. Area of intensive rill formation, **4**. Footpath, **5**. Pasture and rocklands, **6**. Crop land, **7**. Villages.

alluvial plain forming a meandering reach with several point bars and small rapids where the incision has reached the bedrock.

The soils in the area consist of lithosols, fine sandy loams, silty clay loams and clay loams in a sequence from the crest of the mountain to the river. The majority of the soils on the foot-slopes show duplex soil characteristics and have developed deep cracks leading into underground pipes (Fig. 5.2) and are severely gullied. The area is mainly grassland with some small groves of planted trees (poplars and wattle) and willows in the badly gullied areas and along the river banks.

Land use in the area consists of three categories: pasture on the upper slopes, cultivation on low-lying areas along the river plain and on gently sloping parts of the pediment slopes, and villages in the zone between the cultivated and pasture lands.

5.1.2 Catchment Surveys

The area was studied at three levels for the inventory of erosion processes:

1. Air-photo comparison of the degree of erosion and erosion features using aerial photographs taken in 1951, 1961 and 1971.

2. Field surveys during the rainy seasons 1974/75, 1975/76 and 1976/77. This consisted mostly of the inventory of gully conditions and bank erosion observations along the main stream banks and observations of the size and changes of point bars within the Khomo-khoana river banks.

3. Slope transect measurements along some gullies, and cross profiles of the Khomo-khoana river measured in 1975 and 1977 to check gully growth and the shift in the point bars and lateral extension of the actively eroding reaches of the main stream and some main side gullies.

Fig. 5.2. Part of the southern peak of Muela mountain. The landform zones and land use patterns are clearly depicted with the Cave Sandstone plateau in the background, followed by the escarpment and scree slopes, pediment slopes and river valley flats, the central part of which is marked by the line of trees. The foreground shows the pediment slope on the opposite side of the river. (Photo: Q. K. Chakela, March 1975.)

5.1.3 Results and Discussion

In order to simplify the description of the area, it was divided into four zones on the basis of slope gradient, soils and land use. In the following a summary is given of the results of the catchment surveys in the area.

Sandstone Plateau Outlier

This is the highest zone in the area and is represented by three residual hills forming Muela mountain. The hills are flat-topped and underlain by Cave Sandstone formation with a very thin cap-rock of basalt. They are separated by a narrow ridge with sparse vegetation and thin soils on bedrock. The central and northern peaks are small and almost conical in shape with sandstone blocks or a flat surface of sand-

stone at the top (Fig. 5.3) while the southern hill is relatively larger (Figs. 5.1a and 5.2). The cliffy part of the escarpment is concentrated around the southern hill, where there is a series of cliffs along two levels. The vegetation is mainly grasses which are severely overgrazed and some shrubs (*Rhus* spp. and *Chrysocoma* spp.). The soils consist of thin, sandy, stony lithosols derived from the weathering products of the underlying sandstone. The main erosion features in the zone are wind erosion and regolith stripping at the edges of the plateau exposing bare bedrock areas. Raindrop erosion is also active in the zone, as can be seen from the bush and earth pedestals.

Escarpment and Scree slopes

This landform zone extends from 1690 m

113

Fig. 5.3. The northern peaks and slopes of the Muela mountain. Note the bedrock terraces and severely gullied nd surface-washed zone in the lower parts of the slope. This has resulted in abandoning of several fields due to exposure of a hard crustal layer. (Photo: Q. K. Chakela, March 1975.)

to 1750 m and is composed of the following landforms:

a. Steep, stepped or convex escarpment with cliffs in some parts (Fig. 5.2). The southern hill shows a series of cliffs along two lines separated by a minor bench. Around the northern hills the steep slope is reduced both in gradient and length into a series of bedrock scarps separated by minor benches underlain by more resistant sandstones.

b. Scree slopes with old colluvial lobes with boulders (Fig. 5.2). The colluvial lobes vary in size from 50 m to 300 m in width and up to 300 m in length. They are concentrated around the southern hill (Fig. 5.2). These slopes are washed by several ephemeral streams which may flow for the duration of the wet season or just after rain storms.

The main erosional processes observed in this zone are *surface wash* and *rill erosion* on the sides of colluvial lobes. Several boulders have been undermined by weathering and seepage. This has been aggravated by goats and sheep resting under the boulders. Several small sheep-tracks form parallel terraces around the slopes and run parallel to the contour lines. Another significant erosion feature in the zone is the formation of rills along or starting from the footpaths. Around the northern hills several slide scars 20—200 m wide were observed, but these are no longer active and have been completely covered with vegetation. Several rills are found on the northern slopes on cultivated fields on these slopes. Piping occurs in places. Surface wash has advanced so far that most of the fields have been abandoned and a hard soil crust with sparse vegetation cover is the result (Fig. 5.3). The upper parts of the rills and small gullies in this zone are eroding and the eroded materials are deposited within the slopes in the form of small alluvial fans at the mouths of the gullies.

Foot Slopes

This zone starts at the change in steepness of the slope below the scree slopes with the

114

Fig. 5.4. Rills formed on a surface-washed area, area 1, Khomo-khoana catchment. The material washed out of the rills is deposited at the lower limit of the zone of rilling. (Photo: Q. K. Chakela, March 1975.)

upper boundary approximately along the footpath running in an east-west direction and dividing the area into pasture-villages and cultivated land use units (Fig. 5.1b). This zone consists of a rectilinear slope with slight concavity in the lower sections. The gradient of the slope varies from 5° to 20°. The soils in the upper parts often consist of fine sandy loams with abrupt boundaries between the top soils and subsoils along a crustal layer. In the lower parts of the slopes, the soils are deep, black silty to clay loams. The same types of soils occur along the topographic depressions descending the slopes. The upper parts of the zone form the loci of headward extension of most of the gullies in the area and are characterized by bare ground areas around homesteads and surface-washed areas in the interfluves, where this has advanced to formation of rills on the exposed bareground (Fig. 5.4). The gullies are narrow, shallow and often do not reach

the main stream (Fig. 5.5). Some of the gullies continue to the main stream in the form of underground channels (pipes). Very few of the gullies on this slope showed any changes during the study period.

River Plain and Channel Zone

This includes the narrow zone along the Khomo-khoana river and the stream reach within the channel walls. It consists of thick, dark-gray to black deposits which have been incised by the river, forming a meandering reach with alternating sections of point bar formation and bedrock-exposed bottom sections with small rapids. Lateral erosion is very intense and occurs at three locations along the river banks:

a) Along footpaths crossing the main stream.

b) At inflow of side gullies descending from the upper sections of the foot slopes

Fig. 5.5 Discontinuous gullies with small alluvial fans at the mouths. The zone between the gullies is severely affected by surface wash resulting in total removal of the top soil. (Photo: Q. K. Chakela, March 1975).

and along pipes.

c) Where the meandering thalweg has undermined the side banks and in those areas where extensive cracking is dominant even though the main flows never reach the side banks due to protection by point bar deposits.

During the study period, no single side bank had collapsed at the same site more than once.

The highest parts of the area are dominated by overgrazing and consequent wash and wind erosion during the windy months of early spring, and regolith stripping exposing bare bedrock. The escarpment and scree slope zone is dominated by rillwash and minor slumping of the boulders and possibly landsliding during extremely wet years or after extended heavy rains. The footslopes and river plain zones are dominated by surface wash and rill erosion in the upper parts of the slopes where the gradients are relatively steep, rills, and discontinuous and continuous gullies, piping and bank erosion. Deposition occurs at the outlets of some of the short gullies which never reach the main stream in the form of small alluvial fans. Within the river banks the dominant processes are point bar formation, lateral erosion and downcutting exposing the bedrock in several reaches.

The whole area is severely eroded and most of the fields have lost their top soils revealing a hard, crustal layer on which no vegetation seems to have time to establish itself. The subsurface drainage prevalent on the slopes of the area is one of the factors that have led to severe drying up of the slopes as most of the water reaching the slopes in the form of precipitation seeps easily through the cracks into the underground channels and the soil is deprived of moisture storage. This is probably the cause of stunted crops on the slopes as compared to those along the river plain and on the opposite slopes.

116

Fig. 5.6. View over area 2 (Kolojane), Khomo-khoana catchment illustrating the major landform elements and land use pattern in the area. (Photo: Q. K. Chakela, Feb. 1977.)

5.2   Khomo-khoana catchment area 2

5.2.1 Introduction

The centre of this catchment is located at 28° 02′ E and 29° 01′ S within the Khomo-khoana catchment (Chapter 2, Fig. 2.15). It has a maximum relief of 260 m ranging from 1660 m at the confluence with the Khomo-khoana river to 1920 m above m.s.l. at the highest points. The catchment has a northern aspect and the main channel flows to the north. The catchment is crossed by three dolerite dykes, one at the southern water divide and the other two crossing the catchment in N—S and E—W direction about 200 m upstream of the confluence with the Khomo-khoana river. The catchment is a third order catchment. The main channel is perennial but has very low water during the dry period and in most winters only parts of the channel carry any water. The landforms in the catchment consist of Cave Sandstone plateau, Cave Sandstone and Red Beds escarpment, scree slopes, pediments and spurs with an alluvial fill forming the

central part of the catchment (Figs. 5.6, 5.7, & 5.12). The quarternary deposits within the catchment are: alluvial, colluvial and aeolian deposits plus weathering products of dolerite and sandstones.

The soils within the area are mainly fine to very fine sandy loams (West, 1972) with some lithosols on sandstone and dolerite areas. The native vegetation is mainly grasslands which have been modified by cultivation and supplemented with planted wattle trees and poplars as a soil conservation measure and to supplement fuel wood.

5.2.2 Catchment Surveys

The catchment surveys in this catchment comprised the inventory of gully erosion through air-photo comparison between the 1951 air photographs and later, field checking in 1973, 1974/75, 1975/76 and 1976/77 rainy seasons, longitudinal profiling of the main gully and measuring depth, cross-sectional widths in order to check erosion/sedimentation and lateral extension of the gully and headward growth of the side gullies. These data are

Table 5.1. *Observations of Lateral Erosion Along the Main Gully, Area 2, Khomo-khoana Catchment.*

Site	Distance from the Confluence m	Length m	Width m	Depth m	Width of the Section m	Date of observation Month/year
1	141	3.4	2.0	0.5	20.0	12/1974
2	141	1.9	1.6	0.4	31.0	12/1976
3	292	13.1	2.9	11.8	27.0	12/1974
4	1,090	1.7	1.8	3.2	17.0	2/1975
5	1,390	3.7	2.6	4.0	27.0	12/1974
6	2,007	6.7	2.1	6.0	30.0	2/1975
7	2,080	3.4	1.7	6.0	50.0	12/1976
8	2,390	5.6	2.5	4.0	10.0	2/1975
9	2,340	3.9	1.2	4.1	15.0	3/1977

summarized in Tables 5.1 & 6.2 and Figs. 5.6—13.

5.2.3 Results and Discussion

The catchment is divided into four land-form units for the presentation of the study results and the discussion of sedimentation and erosion in the catchment.

Cave Sandstone Plateau

Within this landform unit, which forms the highest parts in the catchment, over-grazing is the dominant feature, leaving only short stands of grass and tussocks of bitter grasses (e.g. *Enionurus argenteus*) with bare soil or rock in between the tus-socks. Bedrock stripping is the dominant erosion type followed by wind and raindrop erosion. The latter two are re-sponsible for the bush and grass pedestals with fine sandy accumulations around these raised mounds. The gradient of the landform unit is flat to gently sloping with the main slope of the surface towards the west. At some points only a ridge of dolerite dyke separates the catchment from the neighbouring catchment (area 3).

Escarpment and Scree slopes

At the headward (southern) side the catchment descends from the Cave Sand-stone Plateau by a vertical bluff of Cave Sandstone capped cliff, concave in plan, and broken at the centre by a dolerite dyke. At the foot of the cliff is a scree slope with two forms. Along the western and eastern fringes of the catchment, the scree slope is steep, short and straight, and in the middle the scree slope is less steep, long and concave upwards. The scree slopes are strewn with rock debris of boulders and fines in which the boulders have been embedded. The grass vegetation is dominated by *Hyparrhenia* species, *Themeda triandra* and some shorter gras-ses. In the centre of the valley-head slopes of the catchment the scree slopes are cov-ered by colluvial deposits which grade to alluvial deposits along the central part of the catchment's lower section. The gra-dient of this landform unit gradually in-creases from 11° in the lower sections to a vertical cliff at the top.

The dominant erosion forms in this landform zone are surface wash on the central part of the valley-head slopes and rilling. The rills form away from footpaths along the slope. These footpaths are in places furrows formed on the exposed bedrock. Rockfall was once an effective process for the retreat of the scarp but at present the activity is very low and no fresh rockfalls have been observed since 1950. The boulders on the scree slopes have been stabilized by the finer materials and vegetation.

Fig. 5.7. The upper reaches of area 2, Khomo-khoana catchment. Note the gullied reach in the centre of the photograph and insignificantly incised reach in the lower parts. (Photo: Q. K. Chakela, Feb. 1977.)

Pediments and Spurs

Below the scree slopes, the central parts of the catchment are flanked by two gently sloping rounded spurs, broad on the zone nearest the scree slopes and narrowing away from the scree slopes. These broad sections of the surface are the remnants of the pediment slope, in which the stream has cut down. The zone is characterized by a rocky surface with a very thin surficial cover. A minor scarp (2—5 m high) separates this upper portion from the rest of the surface, which is covered by a reddish brown to light red soils sequence which consists of red fine sandy loam on the spur crest, pale gray to dark grey very fine sandy loams to silty loams on the spur-side slopes, and dark grey to black silty loams and clays along the centre of the catchment.

The uppermost parts of this landform zone are occupied by homesteads and here footpaths and wash erosion seem to be the dominant erosion forms. Very little rilling is noticed here but all the rills in this zone are along footpaths made by livestock and people converging from all sides towards the villages. Large areas around the homesteads are completely bare of vegetation and during the early spring sands are piled up by wind against bushes and building structures.

The side slopes, which are very intensively cultivated, are dominated by series of rills which seem to last a very short time (a year at most). On aerial photographs these rills are well displayed, but it is very difficult to see them from the ground. Some of these rills have developed into gullies (deeper than 1 m) since 1951 and some have extended to the main channel, particularly in the upper reaches of the landform unit.

The crests of the spurs and areas between the rills and gullies on the spur-flanks are dominated by surface wash. Surface erosion has not advanced to the same level as in area 1, where it has led to some fields being abandoned.

The pediments and spurs are thus dominated by surface wash, livestock trampling and overgrazing, rill wash and to a less degree by gully formation.

119

Fig. 5.8. The lower reaches of area 2, Khomo-khoana catchment. Part of the Khomo-khoana river is seen in the foreground with sand accumulations within the river banks and bedrock outcrops at the confluence with the main stream channel of area 2. Several alluvial terraces can be seen along the Khomo-khoana river. Note also the old side-gully to the left of area 2's main stream channel, this gully has been inactive ever since 1961. (Photo: Q. K. Chakela, Feb. 1977.)

Alluvial Flats

This is a narrow zone along the centre of the catchment. It consists of up to 10 m thick deposits which are entrenched by a branching gully system. The main gully forms a valley-bottom discontinuous gully. The first section, from the confluence to the Kolojane-Hoatane road, is very deep and is cut in the main river alluvium exposing a profile with thick alternating dark and grey horizons (Fig. 5.8). The lowest section of this reach has been incised to bedrock. Several slumps were observed in this reach (Fig. 5.9) throughout the study period, leading to a maximum increase of the gully width by 4 meters. The tendency is to produce a vertical-walled, broad-bottomed cross-sectional profile. This ended below a dolerite dyke near the road. In 1975 the dolerite dyke was removed and a concrete channel was built there with a

concrete bridge (Fig. 5.10). The result was an incision in the reach above the bridge which was a depositional area earlier.

Upstream of the concrete bridge there was a depositional zone, with a meandering channel varying in depth from 0.4—1.65 m on a reach 800 m long. The section is covered by thick fine sandy and silty deposits with gravelly and coarse sandy layers at some points. The old gully walls are found as shallow rounded scarps 10—15 m away from the present narrow channel. The thickness of the deposits in this reach vary from 1 to 2 m and cover a zone 10—15 meters wide and are mostly colonized by grasses. After heavy rainstorms, large amounts of debris (grass and leaves) were observed on these deposits and the whole zone seems to have been flooded with main flow away from the central channel. Very small changes in the chan-

Fig. 5.9. Detail of slumping activity within the lower reach of the main stream channel of area 2. (Photo: Q. K. Chakela, Feb. 1977.)

Fig. 5.10. Concrete culvert and flume on the main stream channel of area 2, Khomo-khoana catchment. The structure has led to downcutting upstream where the channel was aggrading before 1975. The dolerite dyke (at the lower end of the flume) used to be the upper limit of incision of the lower reach of the stream channel. (Photo: Q. K. Chakela, Feb. 1977.)

Fig. 5.11. General view of the gully system of area 2, Khomo-khoana catchment. The discontinuous nature of the gully system is mainly controlled by the underlying bedrock terraces (lower end of the foreground system). (Photo: Q. K. Chakela, Oct. 1973.)

The lower section of this reach is depositional up to a secondary bedrock scarp, 1.75 km from the outlet, where it is freshly cutting through a sandstone/mudstone contact zone. The gullies show no lateral extension in the whole reach, and even the side gullies have grassed floors and sides. The entrenching reaches a maximum 1.8 km from the confluence with the Khomo-khoana river, and the depth decreases to the end of the branched zone upstream where the gully has a flat bottom covered with grass.

The headward continuation of the first order gullies continue into the scree slopes, where two of the gullies have entrenched sections and are actively eroding laterally (Fig. 5.11). These discontinuous gullies are entirely formed in fan-like deposits of stratified colluvium.

The Cave Sandstone plateau is characterized by overgrazing, bedrock stripping, and colonization by Karroid bushes. Limited wind erosion occurs in this landform zone, especially during the windy months of August and September. The escarpment and scree slopes section of the catchment is affected by surface wash, rill and gully erosion. Some of the gullies are formed along livestock tracks and footpaths. Overgrazing is severe on the scree slopes because they form the major grazing zone within the catchment. The pediment and spur slopes zone is marked by biotic erosion (footpaths, resting grounds for animals), rill wash and gully erosion. The latter erosion type is limited to the lowest sections of the slopes. The central, alluvial section of the catchment is both depositional and erosional. It is characterized by a system of valley-bottom gullies with bedrock scarps within the gully walls, erosional in the upper and lower sections and depositional in the middle zone. The gullies have very little lateral activity, and all the headward tributaries end below bedrock scarps with very thin soil cover and thus these gullies have very limited chances for further

nel size were noted in this reach between 1951 and 1971 when comparing the aerial photographs taken in these years. The longitudinal gradient of the whole reach is less than 1°.

The second entrenched zone begins 1 km upstream of the confluence (Fig. 5.7), where the first main tributary channel enters the central gully system. But the depositions are confined to bends within the gully walls which showed no activity during the whole of the study period. The deposits are coarser than in the lower parts of the gully system, but they are still dominated by fine sandy to silty sediments. The sandbanks formed within the gully walls are being stabilized by vegetation and force the thalweg to shift and form meanders along the bottom of the gully.

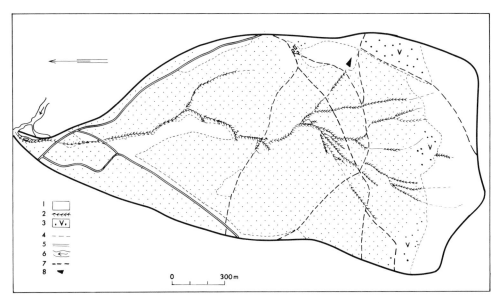

Fig. 5.12. Gully erosion, land-use and communications, area 2, Khomo-khoana catchment. **1**. Pastures and rocklands, **2**. Gullied areas, **3**. Villages, **4**. Non-entrenched stream channel reaches and rills, **5**. Roads, **6**. Khomo-khoana river, **7**. Footpaths, **8**. Dam.

headward extension. The gullies formed above the minor scarps, in the colluvial deposits, extend laterally through slumping without any downcutting. The materials eroded from these and the upper slopes are deposited as small alluvial fans in the zone just above the secondary scarps. The sediments that pass through these depositional zones are deposited in the grassed floor of the main gully, where the reach has a very gentle longitudinal profile. Most of the material is transported during and after heavy rainstorms. Normally the channels have very low water discharges that are insufficient to transport the sediment out of the catchment.

5.3   Khomo-khoana catchment area 3

5.3.1 Introduction

The catchment shares the northern boundary with area 2 (Figs. 5.14 and 5.15). The aspect of the catchment is to the

southwest and descends from the Cave Sandstone escarpment through a long, straight scree slope to the alluvial plain of the Koeneng river (cf. Chapter 2, Fig. 2.15). The catchment studied has an area of 1.1 km². The principal landforms in the area are escarpment, scree slopes and the alluvial plain of the Koeneng river (Figs. 5.15—16, and 5.19). Where the main channel reaches the alluvial plain, a small alluvial fan has been formed by the sediment eroded from the gullied reach extending to the valley-head slopes. Here the channel has been diverted along the contact zone between the scree slope and the alluvial plain towards the west. Several small colluvial fans are located at the foot of the scree where the ephemeral and intermittent streams reach the alluvial plain. The relief of the area is 1680—1920 m above m.s.l. The general gradient of the scree slope is 26° and consists of three parts: (1) a vertical cliff or, steep slope capped by Cave Sandstone, (2) a straight to slightly convex slope with gradients from 8°—35° generally, with some con-

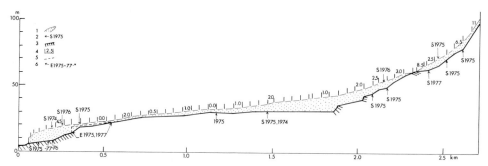

Fig. 5.13. Longitudinal profile of the main stream channel, area 2, Khomo-khoana catchment. **1**. Gullied reach, **2**. Active erosion, type and date of observation, S = Slumping, E = Downcutting, **3**. Bedrock outcrop, **4**. Measured slope section and slope angle in degrees, **5**. 1975 bottom profile within the actively downcutting zone, **6**. Zone of active downcutting due the removal of headscarp in 1975.

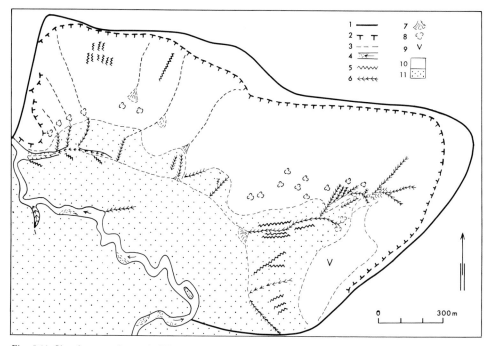

Fig. 5.14. Sketch map of area 3, Khomo-khoana catchment showing land-use and gully erosion based on air-photo interpretation of air-photographs taken in 1961 and 1971. **1**. Catchment boundary, **2**. Bedrock scarps and cliffs, **3**. Intermittent streams and non-entrenched reaches, **4**. Koeneng river with sandbars within the river banks, **5**. Zone of intensive rill formation, **6**. Gullies, **7**. Alluvial/colluvial fans, **8**. Trees, **9**. Village, **10**. Pastures and rocklands, **11**. Crop land.

cavity in the lower section, and (3) a gently-sloping to almost flat foot slope (0°–5°) stretching from the end of the scree slope to the main drainage channel. The main stream within this area is the only intermittent channel, the rest are ephemeral.

Vegetation consists mainly of grasses with some shrubs in the upper portion of the scree slope and at the head of the drainage depressions. Five woodlots are found in the area. These have been planted with poplar and wattle trees along the main gully. In comparison with area 2, this area has a completely different grass vegetation. The *Hyparrhenia* species that dominate the scree slopes in area 2 are

124

Fig. 5.15. Aerial view of area 3, Khomo-khoana catchment. (Les Col 71:2, 5:1579, May 1971.)

limited to north-facing slopes near the village. The main slopes are covered with completely different types of grass vegetation and shrubs are of another species compared with those of the other catchment. The dominant soils are lithosols above the escarpment, very fine sandy loams, sandy loams and colluvial soils in the mid-slope regions grading to fine sandy loams and silty clay loams and clays on the lower parts and along the alluvial plain of the Koeneng river. Due to overgrazing, large parts of the scree slopes have been invaded by Karroid bushes (*Chrysocoma* spp.). Land use and land management practices in the area are the same as for area 1 (see also Chapter 2, Section 2.3.5).

The land use in this catchment (Fig. 5.14) can be divided into:

1. Cultivation along the alluvial plain and most of the gently sloping parts of the scree slopes.

2. Pasture lands on the major part of the scree slopes and between the fields during the summer months and on the Cave Sandstone plateau.

3. Homesteads and woodlot areas.

5.3.2 Catchment Surveys

The same studies were carried out in this area as in areas 1 and 2. In addition, three slope profiles were made, one along the main gully from the confluence with the Koeneng river to the bottom of the escarpment cliff at the head of the main drainage channel (Fig. 5.17, P3), the other two along some side gullies to the main drainage channel to the gully-head and to the bottom of the escarpment (Fig. 5.17, P1 & P2). In connection with the measurement of the longitudinal profile of the main channel, samples were collected of the deposits along the channel (Figs. 5.17 & 5.18). The main gully and the gullies along which the profiles were made were inspected annually and changes within them were recorded.

5.3.3 Results and Discussion

Erosional activity in the highest parts of this area are similar to those described for area 2.

The scree slopes can be divided into two groups:

1. *Valley-head slopes*—semi-circular at the

Fig. 5.16. Major landforms within area 3, Khomo-khoana catchment. From foreground to background these are: Spur-side slopes, alluvial plain, spur-end slopes, major stream channel's alluvial plain with the line of trees marking the location of the Koeneng river, scree slope, escarpment, and sandstone plateau. (Photo: Q. K. Chakela, Feb. 1975.)

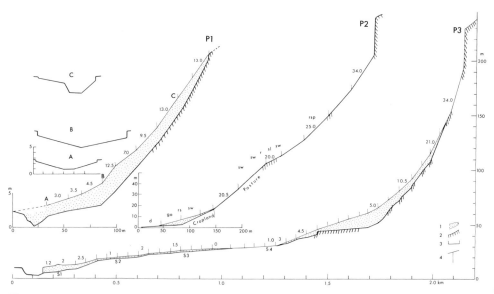

Fig. 5.17. Longitudinal profile of the main stream channel (P3), Slope profile (P2), and a longitudinal and transverse profiles of a side gully (P1), area 3, Khomo-khoana catchment. **1.** Gullied reach, **2.** Bedrock outcrop, **3.** Measured slope section and slope angle in degrees (not all the measured angles are indicated), **4.** Land-use boundary (only on P2), d = deposition, ga = actively eroding gully, r = zone of active rill formation, rsp = zone marked by regolith stripping, sw = surface wash, sl = landslides. S1, S2, S3 and S4 on profile P3 indicate the approximate location of sediment sampling points (see Fig. 5.18).

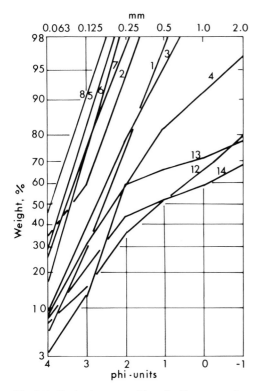

Fig. 5.18. Grain size composition of sediment samples collected from the depositional points along profile P3, Fig 5.17. Samples 1—3 from S1, 4—6 from S2, 7—9 from S3, and samples 12—14 from S4.

heads of drainage channels. They are found in two locations in this area; at the head of the main channel and at the western end of the study area. They consist of concave downward slopes covered with thick colluvial deposits in the lower sections. These deposits have been severely gullied producing gullies up to 10 m deep (Figs. 5.17 & 5.15).

2. *Valley-side slopes*—the open side slope to the main drainage basin (Koeneng Catchment). Here the slope forms are rectilinear with convexities and concavities at the base and they are covered with thin, stony soils. The slope profile, P2 in Fig. 5.17, is measured across a side slope.

Several ephemeral streams descend these slopes and have entrenched lower parts. Several old erosion scars mark the slopes. Although colonized by vegetation

some of these scars and the boulders which have formed them can still be observed (Fig. 5.19). The heads of most of the gullies in the depressions are fed by springs and several pipes open into the main gully north of the village of Matoanes.

The area is severely eroded in the upper, steeper slopes and along the topographic depressions. The erosion types include surface wash, rill- and gully erosion. The lower reaches of the catchment are partly depositional. This has been increased by man-made diversion ditches which have stopped sandy sediments from the gullies from flooding the fields along the alluvial plain. Gully-head extension studies show that several of the gullies which were active in 1951 have now ceased (Fig. 5.14). The highest gully extension rates are confined to the main gully head where three side gullies extended headward more than 100 m between 1951 and 1961. These gullies are reported to have existed as underground channels prior to 1951 and sometime between 1953 and 1960 the roofs of the tunnels collapsed revealing a gully more than a hundred metres long, with depths varying from 5 m at the headscarp to 10 m at outlet to the main gully and 10 m wide at the mouth (Oral information from local inhabitants). The main gully is now cutting through the mudstones and shales in its bed producing a multiple scarped longitudinal profile (Fig. 5.17,P3). The major headward branches of this gully system have reached a bedrock surface. The main channel is discontinuously incised and consists of two entrenched sections separated by a semi-depositional reach starting at the old alluvial fan at the outlet of the gullied reach, where the channel was diverted to the west by a diversion ditch.

Lateral erosion of these gullies is very sporadic, as is the case with most of the gullies inspected in this study. No single gullyside site was observed to undergo sapping more than once during the study period. The collapsed blocks have a

127

Fig. 5.19. Scree slopes and escarpment zone of area 3, Khomo-khoana catchment. Note the fossil landslide scars and scars left by huge boulders on the slope. Some of the boulders are embedded in the lower parts of the slope. Profile 2 on Fig. 5.17 is measured along the gully ending in the tree grove to the right. (Photo: Q. K. Chakela, March 1975.)

tendency of stabilizing the zone just up-stream of the blocks and undermining of lower section through caving by the divided flow.

5.4 Khomo-khoana catchment area 4

5.4.1 Introduction

Khomo-khoana catchment area 4 is similar in form, exposure, landforms and geology to catchment area 2 except for the following: the spurs and pediment zone are very degraded leaving only a bedrock exposed ridge along the western boundary, the side slopes are shorter and steeper and severely gullied to the bedrock. The central part of the catchment consists of stratified deposits forming an in-fill material in a spoon-shaped depression cut in

the bedrock (Figs. 5.20 & 5.25). At the lower end of these mixed alluvial/colluvial deposits is a narrow alluvial zone where the present channel is building small terraces limited by the bedrock bottom exposed in most of the reaches.

The escarpment and scree slope zones are not as marked as in the rest of the catchment. The escarpment consists of steep, stepped slopes with cave-formation along two lines (Figs. 5.20 and 5.21). From above the Cave Sandstone escarpment to the water divide, the slope is convex. The water divide is located in the foothills zone where there is some basalt cap-rock in places.

The soils, vegetation and land use practices in this area are the same as for areas 2 and 3. The larger part of the area was planted with trees and fenced off as a soil conservation measure as early as the beginning of 1950s. The trees have now

Fig. 5.20. Spoon-shaped, wooded hollow of area 4, Khomo-khoana catchment. St Theresa mission to the left. Note the semi-circular valley-head slopes with extensive bedrock outcrops. (Photo: Q. K. Chakela, Feb. 1977.)

Fig. 5.21. Woodlot area within area 4, Khomo-khoana catchment and the Cave Sandstone escarpment in the background. (Photo: Q. K. Chakela, Feb. 1975.)

129

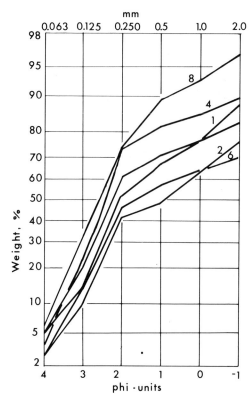

Fig. 5.22. Grain size composition of sediment samples collected from the depositional areas along the main stream channel, area 4, Khomo-khoana catchment. Samples numbers indicate relative locations of the sampling points upstream from the outlet of the catchment.

completely grown and the central part of the catchment is a woodlot area of poplars, spruces and cypresses (Fig. 5.21).

5.4.2 Catchment Surveys

The same methods were used here as for the study areas described previously. In the measurement of the main gully's longitudinal profile, the readings were taken along the gully bottom because trees limited viewing on the gully side. The results of the field work in this area are presented in Figs. 5.22, 5.23 and 5.24. Fig. 5.24 shows the gullies studied by air-photo comparison, location of ground gully-growth sites, location of cross-sections

where repeated measurements were made for lateral extension and growth or decrease of gully bed deposits. In Fig. 5.23 depositional and erosional reaches are indicated. The sites where lateral erosion was observed during the rainy seasons 1974/75, 1975/76 and 1976/77, and the sites of repeated cross-sectional measurements, also marked on Fig. 5.23.

5.4.3 Results and Discussion

Cave Sandstone Plateau

The dominant erosion types in this zone are wash and rill erosion on sloping cultivated areas and limited gullying. The zone grades diffusely into the escarpment zone, through a broad area of thin regolith cover and sparse vegetation and several water springs oozing at the interface between bedrock and regolith or between two sandstone layers. Large areas of bare rock outcrops abound in this zone. These are a result of regolith stripping and livestock activity (overgrazing, trampling by goats and sheep).

Escarpment and Scree Slopes

This region is limited to the concave, valley-head slopes of the catchment, cut through by drainage channels forming minor falls and rapids. The water has cut into the rock along the main channel producing a channel cut in sandstone. Some of the channels pass through the escarpment along dolerite dykes. The stepped form of this zone has influenced erosion activity. The small regolith covered flat, inter-scarp areas are rilled and the regolith has been stripped off. The upper parts of the colluvial/scree slope are the loci of the end of the majority of valley-side and valley-head gullies. Most of the gullies traverse the whole slope to the bottom of the escarpment and have cut through the underlying colluvial deposits, and also down into the bedrock. The transition

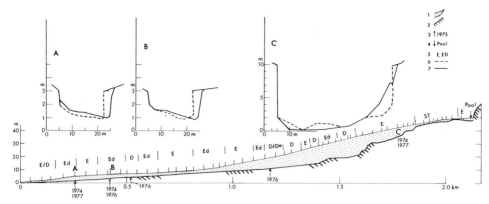

Fig. 5.23. Longitudinal profile and cross sectional profiles of the main stream channel of area 4, Khomo-khoana catchment. The cross sections are located at A, B and C on the longitudinal profile. **1**. Gullied reach, **2**. Bedrock outcrops, **3**. Location and date of observation of active bank erosion, **4**. Location of a plunge pool, **5**. Erosion/deposition zones, **6**. 1975 profile on cross sections, **7**. 1977 profiles on the cross sections.

E = Zone of bed scour and bank erosion, D = zone dominated by deposition, E/D = zone of bank erosion and point bar formation, Ed = dominantly erosional zone with slight deposition, De = dominantly deposition with slight erosion, ST = zone of stabilized gullies due to tree-root binding.

Pediment and Spur Slopes

The pediment and spur zone in this catchment is very much degraded and consists of mainly side slopes to the catchment. The western side slopes are intensively eroded by wash, rill and dis-continuous gullies. Some of these areas are stripped of the soil layer exposing bedrock. In some cases the stripping has been in-itiated through digging of soil for building and plastering of houses. The majority of small gullies and rills, especially in the area outside the woodlot, terminate at the lower parts of the spur-side slopes on the alluvial flat. The interrill/gully areas are affected by surface wash leading to re-moval of the top soil and revealing sandy, pale subsoils. The buffer strips have been transformed into terrace-like forms due to the fact that the sediments washed off from the upper sections of the inter-buffer strip area on cultivated lands are accumulated upstream of the lower buffer strip. The thickness of these accumulations gives a measure of the rate of surface wash on these slopes.

Valley Flats

The central part of the catchment consists of a narrow level to gently sloping zone with a concave upward longitudinal pro-file (Figs. 5.23—25). The zone consists of thick, stratified deposits showing areas of entrenchment and high water-level with several generations of gullying on the side of the present gully walls.

More than 50% of the breadth of the alluvial zone is gullied. There is a dif-ferentiation in grain size composition of the present-day sediment accumulations from the lower parts of the zone to just below the fenced woodlot. The sediments are coarser as one goes upstream from the bridge at the outlet to the catchment to-wards the woodlot. However, the bottom layers in the finer material zones are com-posed of coarse sediments. The inner parts of the deposits are coarser than the outer parts (away from the stream thalweg). Slumping is active in the zone and the channel is being widened.

The upper reaches, within the woodlot

131

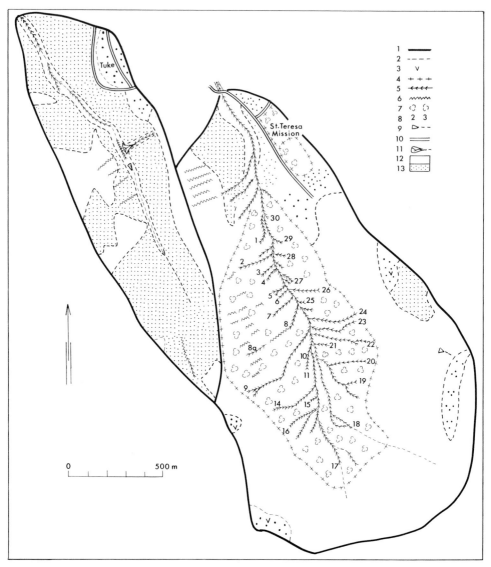

Fig. 5.24. Land use, gully erosion and the location of surveyed gullies, areas 4 and 5, Khomo-khoana catchment. **1.** Catchment boundaries, **2.** Land use boundaries, **3.** Villages, **4.** Fence, **5.** Entrenched stream channel reaches, and side gullies, **6.** Zone of intensive rill formation and sheet erosion, **7.** Trees, **8.** Surveyed gullies, **9.** Dam, **10.** Road, **11.** Fan, **12.** Pastures and rocklands, **13.** Cropland.

area, are severely gullied and several slumps were observed during the study period. In some places the slumps are initiated by the penetration of the root-system of the trees. The bottom of the main gully consists of erosional and depositional sections. The valley-bottom gullies are widening through this section by lateral erosion and the material is washed out by water during periods of high flows.

Table 6.2 summarizes the data on gully-head extension between 1951 and 1971 based on air-photo interpretation and ground checking during the study period. The majority of the gullies show no exten-

132

Fig. 5.25. Aerial view of area 4, Khomo-khoana catchment (Les Col 71:1, 3:1482, May 1971).

sion but all the gullies inspected show widening, especially where they open into the main valley-bottom gully.

The catchment has the highest erosion activity of all the catchments studied within the Khomo-khoana catchment. Gullying is the dominant erosion feature in the central part of the catchment and this has advanced to such a degree that the bedrock is exposed at the bottom of the gullies for a large part of the main drainage channel. The side slopes, both within the woodlot area and on cultivated lands are severely affected by surface wash, rilling and minor gullying. No piping was observed in this catchment. The colluvial slopes are entrenched and the gullies have

reached the bedrock scarp. The escarpment and residual plateau above the escarpment are characterized by surface wash, regolith stripping and rilling along the footpaths. The residual spurs and pediments have been degraded through regolith stripping and surface wash by overland flow and minor gullies.

5.5 Khomo-khoana catchment area 5

5.5.1 Introduction

This small catchment shares the eastern boundary with study area 4 described above. The catchment is completely

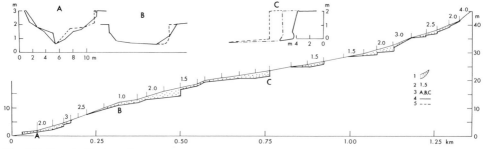

Fig. 5.26. Longitudinal profile and cross sections of the main stream channel of area 5, Khomo-khoana catchment. **1.** Gullied reach, **2.** Slope gradient in degrees, **3.** Location of cross sections, **4.** 1977 profile on the cross sections, **5.** 1975 profile on the cross sections. Note that diagram C shows the headward advance of gully-head scarp measured in 1975, 1976 and 1977.

within the Red Beds Plain and forms a depression on the almost smooth pediment plain. The catchment consists of a single drainage channel which is discontinuously incised (Figs. 5.24 and 5.26). The relief of the area is 125 m ranging from 1660 m above m.s.l. at the confluence with the main stream to 1785 m at the foot of the escarpment. Two dolerite dykes form prominent features within the catchment; one along the eastern boundary running in a N—S direction, the other crosses the area almost in the middle. The lower reach of the catchment, upstream of the confluence, has formed a small terrace within the old gully. Land use, soils, landforms, geology and vegetation in this area are the same as those described in Chapter 2.3 for the lowland region (Red Bed Plains land system).

5.5.2 Catchment Surveys

The catchment surveys done in the catchment are the same as for catchment 1. Fig. 5.26 summarizes the data on longitudinal profile measurements and cross-sectional surveys made in 1974 and 1977.

5.5.3 Results and Discussion

This catchment forms a landform unit often found on the Red Beds Plains land system. The catchment is formed in a single bedrock which seems to have been weathered very deeply in an elongated

bowl and this is filled with fine sandy to silty-clay materials.

The main drainage channel consists of a swampy area in the deepest part of the depression and several springs emerge from the upstream and side slopes of the depression. The main drainage channel is made up of alternating reaches of incision and deposition. This has produced a discontinuous gully system occupying the centre of the depression. One of the side gullies (along a present footpath) has formed a small alluvial fan cut by the entrenched reach of the main gully (Fig. 5.27). The grass vegetation cover along the centre of the catchment is very dense (Fig. 5.28), and it is possible that the sediment accumulation found in the non-entrenched reaches of the main channel are caused mainly by the trap effect of the vegetation.

Fig. 5.26 shows the longitudinal profile of the main drainage channel of the catchment. Two cross-sections were measured in 1974 and again in 1977. These are shown in Fig. 5.26 a—c. Fig. 5.26 c shows the headward advance of one of the secondary scarps of the main channel between 1974 and 1977. The total advance of the scarp is 4 m in 3 years. The greatest gully activity during the study period occurred at two locations along the main gully: (1) at the place where the main stream channel cuts across an alluvial fan of a valley-side gully (Figs. 5.27 & 5.28), and (2) above one of the secondary

Fig. 5.27. Actively eroding reach of the main stream channel of area 5, Khomo-khoana catchment. The main stream channel cuts across a small alluvial fan formed by a side gully initiated along a footpath. (Photo: Q. K. Chakela, March 1977.)

Fig. 5.28. Detail of slumping which occurred after extended rain spell in March, 1977, area 5, Khomo-khoana catchment. (Photo: Q. K. Chakela, March 1977.)

scarps (Fig. 5.26c). In 1977 a slump occurred on the banks of the main stream channel following a rain spell of 14—15 March. The scar left by the slump was 2—3 m wide, up to 11 m long, and 3 m deep (Fig. 5.28). The total widening of the gully in this reach between 1974 and 1977 was measured to be 2 m.

The western slopes of the catchment and the southern (valley-head) slopes show intense rilling and severe wash on cultivated lands. After every storm large enough to produce runoff, large sediment accumulations are found deposited on the

135

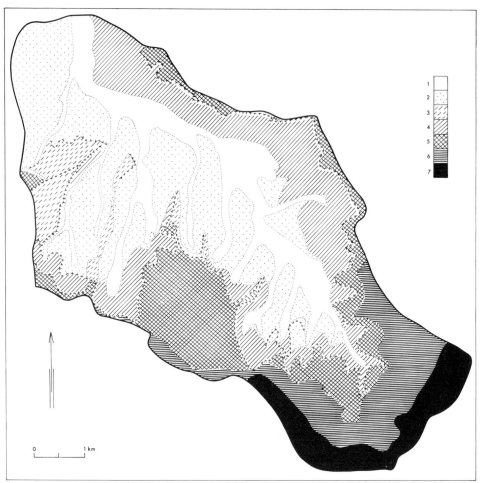

Fig. 5.29. Landform zones, Koeneng river catchment, upstream of Hoatane-Ficksburg crossing, Khomo-khoana catchment. The map is based on air-photo interpretation and field checking in 1975, 1976 and 1977. **1**. River valley flats, **2**. Degraded spurs, **3**. Pediment slopes, **4**. Scree slopes and escarpment, **5**. Cave Sandstone plateau, **6**. Basalt foothills, **7**. Mountain slopes and crests.
Source: Air-photographs over Lesotho: 82E, 1951/52, 57/BS, 1961/62; Les Col 71, 1971.

upslope side of the field boundaries and buffer strips and freshly exposed soils on the downslope parts of the strips. These slopes are thus being successively lowered by rill and surface wash. Where the footpaths cross the main gully at entrenched parts, side gullies are formed. If the gullies or side slope rills reach the centre of the depression at non-incised sections, small depositional forms develop as the sediments brought by storm waters are deposited.

The catchment is not as severely gullied as the other four catchments studied in this area. The side slopes show a high rate of rilling and surface wash similar to that of the rest of the studied subcatchments of the Khomo-khoana catchment. The most typical characteristic of this catchment is its discontinuous gully system. The bottom gullies of areas 2 and 4 are possibly the more advanced forms of the gullying taking place in this catchment at the present.

Fig. 5.30. Major stream channels, gullies, areas of intensive rill formation and wash erosion, and prominent structural forms within the Koeneng river catchment, Khomo-khoana catchment. **1**. Catchment boundary, **2**. Major stream channels and gullies with perennial (solid line) and intermittent streams or non-incised reaches (dashed line). **3**. Zones of dense rill formation and intensive surface wash on spur slopes, **4**. Bedrock scarps and cliffs, **5**. Dolerite ridges, **6**. Alluvial/colluvial fans, **7**. Slide scars.
Source: Air-photographs over Lesotho: 82E, 1951/52, 57/BS, 1961/62; Les Col 71, 1971.

5.6 Erosion and sedimentation within the Khomo-khoana catchment

In this section a general summary of the results of the studies of erosion is presented for the whole catchment. Two simplifications have been made in presenting the results. The Koeneng river subcatchment is used to illustrate the results, and the catchment is divided into five landform zones based on relief,

bedrock, soils, vegetation and hydrology. This division is a simplified form of the main physiographic units of the Khomo-khoana catchment (Fig. 2.17, Fig. 5.29 and Table 2.2).

Mountain Slopes

No significant sedimentation forms exist in this zone, except small alluvial/colluvial fans found at the boundary zone to the foothills region where the gradients be-

Fig. 5.31. Aerial view of the upper reaches of Koeneng river catchment illustrating landform zones (from lower to upper): Pediment slopes, Scree slopes and escarpment zone, Cave Sandstone plateau, Basalt foothills and mountain slopes and crests (57/BS/9: 156, May 1961).

come smaller along the minor drainage channels.

The area is heavily overgrazed and traversed by several footpaths and livestock tracks. Some of these tracks may be as deep as 1 m. The storm waters have helped in deepening these tracks into parallel gullies in the interfluve areas and on minor spurs. The result of the severe overgrazing is pronounced surface wash and rill-formation on the spur slopes. Permanent gullies are almost absent in the zone. The only gullies worth the name in the area are those found at the bottom of the small valleys where the streams are entrenching the wash deposits from the upper slopes (Fig. 5.31).

Foothills Spurs and Cave Sandstone Plateau

The topographic depressions along most of the streams in this zone have local depositional areas consisting of gravels and some pebbles. However, the materials found in these streams are mainly fine textured.

Vegetation conditions indicate severe overgrazing and consequent dominance of surface wash. Most of the cultivated areas in the zone have lost their surface layers and stoney raw materials, soils and weathering products are exposed on the fields. Gullying and rill-formation are confined to valley-head slopes, lower side-slopes and along the alluvial flats of the major streams. The thin soils on the sandstone plateau are affected by regolith stripping resulting in isolated islands of grass-covered patches on rock outcrops.

Fig. 5.32. Aerial view of the middle reaches of Koeneng river catchment illustrating landform zones: Cave Sandstone plateau, (upper left corner), escarpment and scree slopes, river valley flats, with Koeneng river in the centre, degraded spur slopes (two protrusions in the lower part of the picture, separated by non-incised reaches), separated from the valley flats by a minor bedrock scarp. Note also the intensive cultivation and field patterns, and the deeply entrenched reaches of area 3 (to the left of the village) and the main stream channel in the centre. (Les Col 71:4A 2940, May 1971.)

Escarpment and Scree Slopes (Figs. 5.15, 5.32 and 5.33)

The soil cover on the scree slopes is thinner on the upper parts and usually thickens downslope (Fig. 5.16). The slopes are covered with colluvial deposits which form elongated lobes strewn with huge boulders. The valley-head slopes of the depressions and streams emerging below the escarpment are covered with thick, stratified colluvial/alluvial deposits which

Fig. 5.33. Aerial photograph showing the following landforms (from lower to the upper part of the picture): the entrenched alluvial plain of the Futhong river, scree slopes, escarpment, Cave Sandstone plateau, escarpment, scree slopes, pediment slopes (location of villages), lowland spurs separated by gullied topographic depressions (Les Col 71:2, 5:1577, May 1977).

are now being entrenched into deep gullies (Figs. 5.15 and 5.33) with exposure of sub-surface pipes emptying into the main gullies. The major streams pass through these landforms in the form of gorges which continue into the foothills zone. Where the dolerite dykes cross the Cave Sandstone escarpment hollows have been formed and these make the access routes to the sand-

stone outlier plateaux and foothills zone. The dykes continue as ridges below the Cave Sandstone escarpment.

The vegetation cover in this zone consists of tall grasses (*Hyparrhenia* spp.) on north-facing slopes, shrubs and grasses on the south-facing slopes. This differentiation of vegetation indicates the differences in moisture and temperature conditions of

Fig. 5.34. View over part of Khomo-khoana catchment. The landforms depicted are (from foreground to background): Scree slope, river valley flats with the meandering reach of Khomo-khoana river, spur slopes separated by incised tributaries of the Khomo-khoana river, escarpment and mountains. (Photo: Q. K. Chakela, Feb. 1975.)

the slopes with different aspect.

The lower sections of these slopes are normally dominated by deposition of colluvial materials and small fans where the drainage channels reach the gently sloping footslopes (Fig. 5.15).

The dominant erosion feature in these landforms is gullying in the thick colluvial deposits at the valley-head slopes of the drainage depressions (Figs. 5.15 and 5.33). Rill-formation and surface wash dominate the open slopes and interfluve areas. In association with gullies in the colluvial depressions are found several subsurface channels opening into the central gullies. Overgrazing is acute in this zone and has led to the invasion of the slopes by Karroid bushes (*Chrysocoma* spp. and *Aster* spp.).

Pediments and Lowland Spurs

The larger part of this zone is cultivated and the indigeneous vegetation consists of grasses (Figs. 5.32—34). The drainage in this zone consists of intermittent, ephem-

eral and some perennial streams. Several of these streams flow along shallow bedrock depressions and are fed by seasonal spring waters.

The interfluve areas are dominated by surface wash and rill-formation (Figs. 5.30, 5.33). Wind erosion is also active on the open spur ridges. The spur-side slopes are characterized by a high concentration of parallel rills and small gullies (Fig. 5.30) and surface wash on the inter-rill/gully areas. The result of these processes has been removal of the top organic layers on these slopes. The heads of most of the gullies from the inter-spur streams are found in the lower parts of these side-slopes.

River Valley Flats (Alluvial Plain)

The lowest parts of the catchment, along the main river channel and its tributaries, consists of a flat, narrow area with thick, dark to black alluvial deposits (Figs. 5.29, 5.32 and 5.34). Several of the reaches in this zone show terrace formation, point bars (Fig. 5.34) and mid-stream fans. The majority of the fan-like deposits are found on tributary reaches at the points where the tributaries reach the major streams' alluvial plain or where there is an abrupt change of slope of the tributaries' longitudinal profiles (Fig. 5.30).

The major streams have cut meandering channels with steep to almost vertical walls in the alluvial deposits. Within these walls, sandbanks are being formed which seem to be migrating downstream. In certain reaches fans are formed in the mid-main-channel position resulting in reduced downcutting and branching of the main channel into two or more channels. Two such areas have been observed in the subcatchment upstream of Thaba-Lesoba (Fig. 2.15) and an entrenched one within the Koeneng river catchment. These areas either form reed meadows or are bogs with impeded drainage.

The majority of tributary streams are ephemeral channels with very steep-sided, deep gullies with several actively eroding side gullies. The dominant erosion processes in this zone are lateral bank erosion within the major streams and major gullies, gully erosion and piping.

Downcutting is taking place in most of the reaches, but this has advanced so far that the bedrock at the bottom of the deposits is being worn down too. The result is minor potholes and several small rapids and falls along the major streams. Some tributary streams show an upper reach occupied by actively eroding gullies and a depositional reach in the middle (Figs. 5.27 and 5.32). The lowest reach, opening into the major stream, is often erosional depending on the depth of the main stream at the confluence. The typical sequence, beginning at the confluence with the main stream, is: an entrenched zone, a depositional zone, and an actively eroding gullied zone terminating at the foot of the scree slope or below a local bedrock scarp.

The different zones described above are summarized in Fig. 5.29. The extent of gully erosion and the prominent structural and depositional features within the Koeneng subcatchment are presented in Fig. 5.30. These two maps are based on air-photo interpretation of aerial photographs taken in 1961 and complemented in 1971 as regards the lowlands (Figs. 5.1a, 5.15, 5.25, 5.32 and 5.33).

6. Summary and discussion

The purpose of this study was to document the types, extent and rates of erosion and sedimentation in two catchment areas in Lesotho.

6.1 Rates of processes

The rates of erosion and sedimentation were investigated to differing degrees in different areas, dependent on the methods employed and the intensity to which the methods could be used. The studies within the Roma Valley and Maliele catchments are relatively quantitative, because of the availability of reservoirs and the intensity of the water- and sediment sampling during the last field period. However, the data on which these quantitative estimates are based are rather preliminary because they cover a short time-span. Table 6.1 shows the rates of erosion and sedimentation within the Roma Valley and Maliele catchments. The rates of net erosion vary considerably from area to area and cover a range of about 100—2000 t km^{-2} y^{-1}. These values were measured in reservoirs and selected river stations and therefore are only indirect indices of the real rates of erosion in the catchments. Large amounts of eroded materials are deposited upstream of the measuring sites.

Table 6.1. *Rates of Erosion and Sedimentation, Roma Valley, Maliele Catchment and Little Caledon Catchment.*

Catchment	Relief Ratio m/km	Catchment Area km^2	Sediment Yield t km^{-2} y^{-1}	Remarks
Roma Valley area 1	860/7.9	26.0	380	Based on water and sediment discharge measurements, 1976/77 rainy season.
			340	Based on reservoir surveys between 1973 and 1975.
Roma Valley area 2	200/0.8	0.5	1,700—1,870	Based on reservoir surveys between 1973 and 1977.
Roma Valley area 3	50/0.9	0.6		Based on reservoir surveys. The reservoir had such a low rate of sedimentation that it could not be used for sediment yield estimate with any certainty.
Roma Valley area 4	310/2.6	2.2	825	Based on reservoir surveys between 1973 and 1977.
Roma Valley area 5	920/13	57.0	1,370	Based on water and sediment discharge in 1976/77 rainy season.
Maliele area 1	210/1.4	0.6	220	Based on reservoir surveys between 1974 and 1977.
Maliele area 2	325/4.2	6.3	350	Based on reservoir surveys between 1974 and 1977.
Maliele catchment	375/6.5	13.5	270	Based on water and sediment discharge during 1976/77 rainy season.
Little Caledon Basin	1,495/43	945	1,979	After Jacobi 1977.

The catchment surveys and air-photo studies (Table 4.3, Figs. 4.1, 4.16 & 4.38) indicated that the severity of erosion in the catchments varied from one landform zone to another. The spurs and pediment slopes, both in the basalt and sandstone terrain, are severely affected by surface wash and rill formation. Gullies are limited to narrow zones in the drainage depressions and to floodplains of the major streams and their tributaries. The severity of erosion is further aggravated by intensive cultivation of the spurs and pediment slopes in the lowlands and by overgrazing on the steep slopes in the mountains and on the scree slopes. Land use, therefore, is also a major factor in the erosion processes in the studied areas. A clear example of the effect of cultivation on erosion and sedimentation is that large areas with fields and villages show higher sediment delivery to the catchment streams than smaller areas with limited cropland. The comparison of sediment concentration at two Roma Valley river stations during the rainy seasons of 1975/76 and 1976/77 showed increased sediment concentration downstream (Fig. 4.7 & Tables 4.1 & 4.7). The intervening area between the two stations is heavily cultivated and under rapid urban development. The severity of the processes within the Roma Valley and Maliele catchments can be ranked as follows, beginning with the most severe: (1) surface wash and rill formation, (2) gully erosion, (3) local sedimentation at the foot of steep slopes and within gullies, (4) wind erosion on the open spurs and pediments, and (5) mass movements on south-facing slopes.

The greatest erosion caused by running water occurs mainly during two periods: after the first heavy rains following the dry season, and during the time of ploughing and weed clearing. In the former case there is practically no vegetation to intercept the high intensity raindrops, as the cultivated areas are bare of crops and the pastures have not yet recovered their grass cover after the dry season. Wind erosion is also very effective during the early spring (August and September). The lack or shortage of vegetation cover at the beginning of the rainy season and the loosening of the soils on the cultivated lands during the ploughing and weed-clearing periods, combined with the high intensity of the thunderstorm raindrops, all work to aggravate the effect of raindrop erosion and wind erosion during the early, windy part of the spring.

The rates of reservoir sedimentation indicate that most of the reservoirs (exceptions being regulated reservoirs or those with other reservoirs upstream) have a very short useful life. Most of the reservoirs within the study areas are filled within 10 years, and all are filled within 30 years. Reservoir capacity losses are about 4—20% per year. The highest losses occur in the first few years after dam closure. The sedimentation rate slackens as the reservoir trap efficiency decreases.

Within the Khomo-khoana catchment, ground measurement of freshly eroded gully sites and measured sediment accumulation along gully cross-sections indicate an average of 0.1—1.0 mm lowering of the ground surface per year in the severely eroded catchments.

The worst and most wide-spread erosion feature in the Khomo-khoana catchment, though not obvious at first glance, is the loss of surface layer of soils on spur-side slopes and pediment slopes in the lowlands. This leads to bedrock exposure or to the formation of small rills descending the side slopes through cultivated lands (Fig. 5.30). Gully erosion is the most obvious and striking erosion feature. Rates of gully growth, however, are erratic (Tables 5.1, 6.2, and Figs. 5.1, 5.15, 5.25 and 5.30). Rates of gully growth as high as 100 m in ten years have been observed. The main reason for the slow rates of growth seems to be that most gullies have reached bedrock scarps, water divides or zones with very shallow soils, and there is no

Table 6.2. *Summary of Gully Extension Within Some Selected Catchment Areas in Khomo-khoana Catchment Between 1951 and 1977 Based on Air-photo Interpretation and Ground Measurements.*

Area	Gully growth in meters between 1951 and 1977							
	0—5	5—10	10—20	20—30	30—40	40—50	>50	Total
1	6	2	2	2	2	0	2	16
2	21	3	5	1	0	0	0	30
3	11	2	2	1	0	0	3	19
4	21	3	2	3	1	0	0	30
	59	10	11	7	3	0	5	95

further ground for growth. Another factor may be that all areas liable to gully erosion have already been gullied. This is partly supported by the lack of newly-gullied areas since 1951. The lateral extension of several major gullies is limited to the steep upper reaches of the catchments. The lower reaches seem to have stabilized, with deposition along the gully floors. These lower reaches are characterized by stable, shallow walls. Table 6.3 summarizes the severity of different erosion processes from one drainage basin to another within the Khomo-khoana catchment.

6.2 Geomorphological conclusions

The rapid rates of erosion in the 11 investigated areas and those obtained by the Department of Hydrological and Meteorological Services of Lesotho correspond to sediment yields of 100—1900 t km^{-2} y^{-1}. These figures, however, are for very short periods, and the range may be lower for long-term rates of denudation. The longest period of observation is 6 years (Little Caledon). The erosion within these 11 areas leads to rapid sedimentation of (a) sandy, infertile materials at the base of the scree slopes and along gently sloping parts of the inter-spur depressions and (b) silty to clayey sediments on the floodplains of major streams and in the reservoirs. Some arable floodplain land therefore is lost because of such sedimentation. Other results of the erosion and sedimentation

are (a) rapid siltation of several of the reservoirs constructed since the late 1950s, (b) high sediment loads in the major streams, and (c) development of river terraces.

Raindrop erosion, in the form of splash and surface wash (including rill formation), is the most severe erosion process on grazing and cultivated land, both in the basalt and in the sandstone terrain, on spurs and pediment slopes with 3°—6° gradients. Observations of splash pillars, buffer strip heights above the field surface, and of pegs along some of the slope transects show that the average annual rates of surface lowering may be as high as 10—15 mm.

Gully erosion is second in importance and is the most spectacular erosion feature in all the investigated areas. Field observations during this study and air-photo comparison revealed no newly-gullied areas since 1951. The largest gullies are found along the major drainage channels and form headward extensions of these. Discontinuous, valley-side gullies are not wide-spread, but where they occur they are the only gullies that do not follow the general drainage pattern within the catchments.

The third most important erosion feature is the parallel gullies and rills along the livestock tracks. These are formed by livestock and occur between pastures and villages and on large areas of bare ground around the homesteads. Channel erosion, mass movements, pipe erosion and regolith stripping form the fourth group of

Table 6.3. *Qualitative Intensity of Different Erosion and Sedimentation Processes in Five Small Catchment Areas Within Khomo-khoana Catchment.*

Process	Catchment areas				
	1	2	3	4	5
Surface wash	+++	++	++	+++	+
Rilling	+++	++	++	+++	+++
Gullying	++	++	++	+++	+
Piping	+++	+	++	+	0
Local deposition	++	++	++	++	+
Bank-erosion	++	++	++	+++	++
Overgrazing	++++	+++	+++	++++	++

Note: +++++   Extremely severe
    ++++   Very severe
     +++   Severe
      ++   Moderate
       +   Slight
       0   Not observed

active geomorphic processes. These are, however, very local. Wind erosion may be an important process on the open spur and pediment slopes. Some of the bush pedestals may be a combined effect of splash and wind erosion. The maximum wind activity probably occurs during the dry, windy months of August and September, when the cultivated lands are bare of crops and stub, and the pastures have not yet regained their grass cover.

The major geomorphic conclusions that can be drawn are that raindrop erosion and surface wash are more important present-day processes than gully- and channel erosion. The period of maximum gully activity is past; however, gullies have created channels for easy and rapid concentration of runoff and transportation of water and sediment from the catchments.

6.3 Conservation control measures and future research

The high rates of erosion reported are only a quantification of what had been expressed in several reports dealing with land resources of Lesotho (Pim, 1935, Stapples and Hudson, 1938, Morse, 1960, Bawden and Carroll, 1968, Binnie and Partners, 1971 & 1972) and in the internal reports of the Soil Conservation Unit of the Ministry of Agriculture. The main difference is that my study has supplied some quantitative data on small, specific areas. Several conservation measures were recommended in reports dating as early as 1876 (Germond, 1967), and in the present study the effects of four of these measures were studied. These four major conservation measures were mechanically-constructed contour furrows (terraces), grass strips at 2 m intervals (buffer strips), tree plantation, and dam construction across some gullies. In addition, I used rock-walls and rockfills constructed in some areas to check the advance of head-scarps and to encourage deposition within the gullies.

The effect of these measures seems to have been minimal in all the areas investigated in this study. A very noticeable problem is the lack of maintenance of the structures. Some of the contour furrows empty directly into active gullies, with the result that such furrows form loci of initiation of side gullies. The spacing of the buffer strips has not been strictly followed, and some strips have been removed. As a result, erosion from the waters which accumulated on the upstream parts of the strips intensified. Rills and gullies then formed where the water broke through. Most of the rockwalls and rockfills were constructed without considering the morphological properties of the soil profiles and the flow conditions in them. Consequently, such rockwalls and rockfills have been left standing as walls as the gully-head scarp advances by piping or sapping. Some reservoirs (reservoir 2 is the best example) have managed to stabilize gullies both downstream and upstream of the dam. Others have had a limited effect on gullies, even upstream of the reservoirs. Still others, typified by reservoir 6, have led to both increased erosion and to gully retreat downstream of the reservoirs.

It is evident that the need for soil and water conservation measures is urgent and vast in Lesotho. One question which is still unanswered is why the maintenance of the present conservation measures is so bad. Is the problem of erosion understood by the authorities? If so, how is this understanding disseminated to the peasants? A related question is whether the peasant realizes that soil erosion is a problem that can be controlled through the measures that have been implemented. Fundamental to the solution of any problem is the understanding that the problem exists and that something can be done about it.

On the whole, the same recommendations that have been made in other areas severely affected by erosion can be recommended for Lesotho (Rapp et al., 1972, FAO, 1965, Hudson, 1971):

1. Soil and water management in the catchments should be improved. Measures to use: controlled stock numbers and grazing; re-seeding of grass to cover eroded lands; and within gullies, increased planting of grasses and herbs, instead of the trees presently being planted.

2. More moisture should be conserved on cultivated lands. This can be done by mulch cover and by ridging and tie-ridging. On maize and sorghum fields interrow cultivation can be used. This would minimize the channelling of storm waters along the interrow spaces after heavy thunderstorms.

3. Planning and construction of new reservoirs should include estimates of rates of sediment delivery to the reservoirs. Also, detailed maps and capacity curves of the reservoirs should be made immediately after completion of the reservoirs. This will facilitate future studies of reservoir sedimentation. In this way the reservoirs will help estimate rates of erosion in the catchment and thus supply information for checking the effectiveness of the anti-erosion measures used within the catchment.

4. A permanent erosion and sedimentation monitoring system should be established. This should supply information on the rates of erosion and sedimentation in different areas of the country and should encourage research on erosion- and sedimentation control.

For soil and water management improvement in Lesotho, there is therefore an urgent need for continued research into climatic and hydrological patterns. Within Lesotho, only very limited knowledge and useful data are available at present. Such a research programme should also investigate land use, soil loss and sedimentation of different scales within selected catchment areas, ranging from plot studies to studies on catchment scale. In areas without reservoirs, gauging stations should be installed in well situated areas where the gauge readings and protection of the instruments can be assured. Most of my field observation bench marks were ruined by vandals. The best possible results would come from establishing several instrumented catchments in the vicinity of agricultural extension stations, Agricultural Colleges, the National University of Lesotho and within each of the rural development project areas. On the catchment scale the methods used in this study and those in Tanzania (Rapp et al., 1972) could be used with improved density of slope surveys, instrumentation of the river stations and introduction of a dense network of rainfall stations, including measurements of evaporation and temperature. The implementation of such a programme could be taken as a joint venture by the Hydrological Survey's section of the Ministry of Water, Energy and Mining, the Soil Conservation Division of the Ministry of Agriculture, the Agricultural College, and the National University of Lesotho Departments of Geography and Biology. Such a programme would not only supply urgently needed information but would also help in training the even more urgently needed personnel in studies of the problems of erosion and sedimentation. This in turn would strengthen the almost non-existent research and know-how which presently characterizes these institutions.

7 References

Arnoldus, H. M. J., 1977: Predicting soil loss due to sheet wash and rill erosion. *FAO Soil Conservation Guide* 1, 99—124.

Ashida, K., 1980: How to predict reservoir sedimentation. *International Symposium on river sedimentation.* Beijing.

Axelsson, V., 1967: The Laitaure delta: A study of deltaic morphology and processes. *Geogr. Ann.* 49A, 1—127.

— 1979: Sjöars sediment studeras med röntgenteknik. *Ymer 1979*, 133—144.

Bawden, M. G. and Carroll, D. M., 1968: The land resources of Lesotho. *Land Resour. Div. Dir. Overseas Surv. Land Resour. Stud.* No. 3.

Binnie and Partners, 1972: *Lesotho: Study on water resources development.* Inventory Reports 0, 1, 2 and 3. UNDP (Special Fund). International Bank for Reconstruction and Development. Maseru and London.

Bolline, A., 1978: Study of importance of splash and wash on cultivated loamy soils of Hesbaye (Belgium). *Earth Surface Processes* 3, 71—84.

Brice, J. C., 1966: Erosion and deposition in the loess-mantled Great Plains, Medicine Creek Drainage Basin Nebraska. *U.S. Geol. Surv. Prof. Paper* 352—H.

Bryan, R. B., 1969: The relative erodibility of soils developed in the Peak District of Derbyshire. *Geogr. Ann. 51A*, 145—159.

— 1974: Water erosion by splash and wash and the relative erodibility of Albertan soils. *Geogr. Ann. 56A*, 159—175.

Carroll, D. M. and Bascomb, C. L. 1967: Notes on soils of Lesotho. *Land Resour. Div. Dir. Overseas Surv. Tech. Bull. No. 1.*

Carroll, P. H., Nielson, E. C., Howard, R. F., Bishop, W. D., Lepele, J. M., and Nkalai, D. M. T., 1976: *Soil survey of the Thaba-Bosiu project, Lesotho (Technical Appendix).* The Office of Soil Survey, the Conservation Division, Ministry of Agriculture. Lesotho.

Chakela, Q. K., 1973: Water and soil resources of Lesotho, 1935—1970: A review and bibliography. *SIES Report* 2. Stockholm.

— 1974: Studies of soil erosion and reservoir sedimentation in Lesotho. *UNGI Rapport* 34, 479—495.

— 1975: Erosion and sedimentation in some selected catchment areas in Lesotho, 1974/

75. *Dep. Phys. Geogr. Univ. Uppsala* (Stenciled report).

— 1980: Reservoir sedimentation within Roma Valley and Maliele catchments in Lesotho. *Geogr. Ann. 62A*, 157—169.

Chow, V. T. (Ed.), 1964: *Handbook of Applied Hydrology.* McGraw-Hill Book Company, Inc, New York.

Cooke, R. U. and Doornkamp, J. C., 1974: *Geomorphology in Environmental Management.* Clarendon Press. Oxford.

Cooke, R. U. and Reeves, R. W., 1976: *Arroyos and Environmental Change in the American Southwest.* Clarendon Press. Oxford.

Cox, K. G. and Hornung, G., 1966: The petrology of Karroo basalts of Basutoland. *Amer. Min. 51:2*, 1414—1432.

Dempster, A. N. and Richard, R., 1973: Regional geology and structure. In: Nixon, P. H. (ed.), *Lesotho Kimberlites.* 1—19. L.N.D.C. Maseru.

Faniran, A., and Areola, O., 1978: *Essentials of Soil Study with Special Reference to Tropical Areas.* Heinemann, London.

FAO, 1965: *Soil Erosion by Water; some measures of its control on cultivated lands.* Rome.

— 1977: Khomo-khoana rural development project, Lesotho. *Second Meteorological Report. FAO/TF/Les. 9.* Maputsoe.

Flannery, R. D., 1976: *Handbook for Gully Control and Reclamation.* Lesotho Agricultural College. Maseru.

— 1977: Soil erosion phenomena in Lesotho. Lesotho Agricultural College. Maseru. (Stenciled report).

Folk, R. L. and Ward, W. C., 1957: Brazos river bar: a study in the significance of grain size parameters. *Journal of Sedimentary Petrology 27:1*, 3—26.

Germond, R. C., 1967: *Chronicles of Basutoland.* Morija Sesuto Book Depot. Morija. 583 p.

Gill, M. A., 1979: Sedimentation and useful life of reservoirs. *J. of Hydr. 44*, 89—95.

Hudson, N., 1971: *Soil Conservation.* Batsford. London.

Jacobi, S., 1977: *Sediment load estimates of rivers in Lesotho.* Department of Hydrological and Meteorological Services Branch. Ministry of Water, Energy and Mining. Maseru.

Karaushev, A. V., 1966: The silting of small

reservoirs and ponds—theory and calculation method. *Soviet Hydrology: Selected Paper No. 1*, 1966, 35—46.

King, L. C., 1951: *South African Scenery*. Oliver & Boyd. Edinburgh.

Ministry of Water, Energy and Mining, 1970: *Hydrological Data to September 1970*.

— 1970: *Meteorological Data to September 1970*.

— 1975: *Hydrological Yearbook:* from October 1970 to December 1974.

— *Climatological Bulletin*, 1974—1977.

Morgan, R. P. C., 1978: Field studies of rainsplash erosion. *Earth Surface Processes, 3*, 295—299.

— 1979: *Soil Erosion*. Longman. London.

Morse, C. (Chairman), 1960: *Basutoland, Bechuanaland Protectorate and Swaziland. Report of an Economic Mission, 1959*. H.M.S.O. London.

Nilsson, B., 1969: Development of a depth-integrating water-sampler. *UNGI Rapport 2*.

— 1974: Ett exempel på vattenregleringarnas inverkan på sedimenttransporten. *UNGI Rapport 34*, 583—597.

— 1976: The influence of man's activities in rivers on sediment transport. *Nordic Hydrology 7*, 145—160.

Pim, A. W., 1935: *Financial and Economic Position of Basutoland*. H.M.S.O. London.

Rapp, A., Berry, L., and Temple, P., (eds.), 1972: Studies of soil erosion and sedimentation in Tanzania. *Geogr. Ann. 54A:3—4*.

Schmitz, G., 1980: A rural development project for erosion control in Lesotho. *ITC-Journal*, 1980—2, 349—363.

Stapples, R. R. and Hudson, W. K., 1938: *An Ecological Survey of the Mountain Area of Basutoland*. Crown Agents. London.

Stocking, M. A. and Elwell, H. A., 1976: Rainfall erosivity over Rhodesia. *Trans. Inst. Brit. Geogrs. New Series 1*, 231—245.

Stocking, M. A., 1978: The measurement, use and relevance of rainfall energy investigations into soil erosion. *Z. Geomorph. Suppl-Bd 29*, 141—150.

Stockley, G. M., 1947: *Report on the Geology of Basutoland*. Government Printer. Maseru.

Strand, R. I., 1977: *Design of small dams*. Reservoir sedimentation. U.S. Bureau of Reclamation, Water Resources Technical Publications, 767—796.

Sundborg, Å., 1956: The River Klarälven. A study of fluvial processes. *Geogr. Ann.* 38:2—3.

— 1958: A method for estimating the sedimentation of suspended material. *Ext. Compt. Rend. Rapp., Ass. Gén. Toronto, Tome I.* (Gentbrugge), 249—259.

— 1963: Luleälven. Prognos rörande transporten av suspenderat material nedströms blivande dammläge vid Vittjärvi. *Dep. Phys. Geogr. Univ. Uppsala* (Stenciled report).

— 1964a: Sedimentation in and erosion downstream of Tabqa reservoir. *Vattenbyggnadsbyrån*. Stockholm. (Stenciled report).

— 1964b: The importance of the sediment problem in the technical and economic development of river basins. *Ann. Acad. Regiae Sc. Uppsaliensis*. 8, 33—52.

Ward, R. C., 1975: *Principles of Hydrology*. McGraw-Hill. London.

West, B. G., 1972: Soil survey of the Khomo-khoana catchment Basin. *FAO. AGG, SF/ Les. 2. Technical Document 3*. Leribe.

Wischmeier, W. H. and Smith, D. D., 1958: Rainfall energy and its relationship to soil loss. *Trans. Amer. Geophys. Union, 39:2*, 285—291.

— 1965: *Predicting rainfall-erosion losses from cropland area east of the Rocky Mountains*. U.S. Department of Agriculture, Handbook 282.

Wischmeier, W. H., 1976: Use and misuse of the Universal Soil Loss Equation. *J. Soil and Water Conservation, 31:1*, 5—9.

Young, A., 1972: *Slopes*. Oliver & Boyd. Edinburgh.

Young, K. K., 1976: Erosion potential of soils. *Proceedings of the 3rd Federal Inter-Agency Sedimentation Conference*. PB-245 100, Denver, Colorado.